Noise Control of Hydraulic Machinery

FLUID POWER AND CONTROL

A Series of Textbooks and Reference Books

1. Hydraulic Pumps and Motors: Selection and Application for Hydraulic Power Control Systems, *by Raymond P. Lambeck*
2. Designing Pneumatic Control Circuits: Efficient Techniques for Practical Application, *by Bruce E. McCord*
3. Fluid Power Troubleshooting, *by Anton H. Hehn*
4. Hydraulic Valves and Controls: Selection and Application, *by John J. Pippenger*
5. Fluid Power Design Handbook, *by Frank Yeaple*
6. Industrial Pneumatic Control, *by Z. J. Lansky and Lawrence F. Schrader, Jr.*
7. Controlling Electrohydraulic Systems, *by Wayne Anderson*
8. Noise Control of Hydraulic Machinery, *by Stan Skaistis*

Noise Control of Hydraulic Machinery

Stan Skaistis
Consultant
Product Noise and Strength
Birmingham, Michigan

CRC Press
Taylor & Francis Group
Boca Raton London New York

CRC Press is an imprint of the
Taylor & Francis Group, an **informa** business

First published 1988 by Marcel Dekker Inc.

Published 2019 by CRC Press
Taylor & Francis Group
6000 Broken Sound Parkway NW, Suite 300
Boca Raton, FL 33487-2742

First issued in paperback 2019

ISBN-13: 978-0-367-45129-5 (pbk)
ISBN-13: 978-0-8247-7934-4 (hbk)

Visit the Taylor & Francis Web site at
http://www.taylorandfrancis.com

and the CRC Press Web site at
http://www.crcpress.com

Library of Congress Cataloging-in-Publication Data

Skaistis, Stan
Noise control of hydraulic machinery.

(Fluid power and control; 8)
Bibliography: p.
Includes index.
1. Hydraulic machinery—Noise. I. Title. II. Series.
TJ840.S596 1988 621.2'6 88-7084
ISBN 0-8247-7934-7

Preface

This book is derived from over 20 years of research at Vickers Inc., a major hydraulic component manufacturer, and work with its customers who had a need to produce quiet machines. It is directed toward two groups that might be responsible for producing quiet hydraulic components or hydraulically powered machines. One group would include those with a good mechanical engineering and hydraulics background but no familiarity with such subjects as acoustics, hydraulic noise, cavitation, vibration control, or fluid pulsation mechanics. Readers in the other group would have an acoustics rather than a hydraulics background, so I have included somewhat more hydraulics than would be needed by practitioners in this field. This second group may also find R. P. Lambeck's *Hydraulic Pumps and Motors* (Marcel Dekker, 1983) useful in their work.

My experiences with machine designers showed that many already had the knowledge needed for effective noise reduction. Their unfamiliarity with acoustics terminology hid this fact from them, however, so they searched for expertise. In an effort to be of assistance, Vickers published a booklet on acoustic terminology and noise measurement basics. This booklet, "Sound Advice," was so well received that Vickers published another booklet, "More Sound Advice," that also provided some hydraulic noise fundamentals. This second booklet has gone through many printings and is still available from Vickers.

Designers and engineers who were successful in developing quiet machines reported that these booklets helped them feel like they "knew what they were doing" in noise control work. Their comments are the inspiration for this book.

A largely experimental approach is advocated here. The emphasis is on developing an intuitive or qualitative background, rather than quantitative procedures. The use of mathematics has not been avoided, but minimal fluency is needed to apply effectively the concepts presented. Some equations appear, for example, just to reinforce ideas that are stated in words. In some cases, brief mathematical analyses are used to provide a better understanding of basic mechanisms than words can provide, but the mathematical steps are explained in the process. Some analyses are also included for guidance to those who want to do further research, but these are not needed for practical noise reduction work.

Preference is given to the use of English or ft-lb-sec units, which are the most familiar to mechanical engineers. However, since the SI system is customary in acoustics, it is used in many sections so that they correlate with other literature. The SI units used in this book are given in Appendix A along with their English system equivalents. The system is rigorously standardized so rules for writing, abbreviating, and prefixing SI units are also included.

Although it is good for the reader to be familiar with these units, all instruments and the literature are consistent in their use so there is little chance of using the wrong units and making serious errors. In a few cases, when comparisons with more familiar parameters seem desirable, units in the ft-lb-sec engineering system are also given.

Only those topics found to be useful in practical noise control, based on experience with people who produced quiet machines, are included. The goal is to discuss relevant hydraulic, acoustic, and isolation mechanisms in conventional mechanical terminology so that the terms unique to these subjects do not have to be learned beforehand. However, some of these specialized terms are also introduced and used because they are helpful if further study of these subjects proves necessary. Additional references are also included to assist in further studies.

I want to thank my colleagues at Vickers for their support through the years and to acknowledge the help of Ron Becker, Bob Stephens, and Roy Taylor while I was writing this book. I am also grateful to Paul Durocher and Don Sari of the same organization for their generous assistance in providing the illustrations.

Stan Skaistis

Contents

Noise Control of Hydraulic Machinery

1
Introduction

1.1 FOCUS

With few exceptions, noise control in hydraulically powered machines is concerned with conforming to government regulations designed to conserve hearing. In some cases, noise control may be required by competitive pressure brought on by a quieter machine or the need to operate in a quiet environment.

Machines generally have a number of noise sources. One, perhaps two, produce noise that predominates, so that eliminating all other noise sources has little or no effect on the machine's total noise. Therefore, to quiet a machine it is necessary to attenuate the leading noise.

Because there are a large number of noise sources, considerable time and money can be expended on studying and reducing those that do not have much effect on the total. Economics requires that all quieting efforts be focused on identifying and controlling the key noise generator.

Quieting is accomplished by engineering, not by invention. It is generally wasteful of time and money to look for a device or "bold new approach" that will cure a noise problem. Most solutions come from careful analysis of details and making appropriate adjustments.

1.2 HYDRAULIC QUIETING HISTORY

There are many reasons for the rapid growth in the use of hydraulics in the period following World War II. One important reason was that hydraulic power transmission was quieter than the gears, cams, connecting rods, and cables that it replaced. However, as economics led to higher pressures and speeds, it also led to less beefy components. Power levels of machines also grew rapidly. All of these factors increased noise.

By the late 1950s the noise of hydraulically powered machines had risen to a level where they were the subject of frequent complaints. There was also a growth in the public's awareness of noise, fanned by magazine articles (1–3) with titles such as "Is Noise Driving You Crazy?" Further, Industry was concerned about workers suing their employers for hearing losses. The hydraulic industry, fearful that rising noise levels and growing sensitivity to noise would halt its growth, began work on noise control.

Although some work had already been done on quieting power steering systems in automobiles, on an ad hoc basis, concerted efforts to quiet hydraulics started in the early 1960s.

By 1969 the industry's abatement efforts made hydraulic pump noise data commonly available. However, each manufacturer had its own measuring methods, and comparisons of data from different sources were risky. The National Fluid Power Association (NFPA) recognized this problem and organized its T2.7 Sound Measurement Coordination Committee. In 1970 this group provided a standard for measuring pump noise (4). In 1971 it provided a very similar standard for hydraulic motor noise measurement (5). These standards later evolved into International (ISO) Standards (6,7).

1.3 LEGAL LIMITS

The federal government's efforts to reduce workplace noise levels lagged behind those in the hydraulics industry. In 1969 the federal government responded to growing public concern about noise, and new regulations were promulgated for the Walsh-Healy Act. This act, which was passed in 1936, required public contractors to provide good working conditions but did not define what was good. The new regulations set work-space noise limits of 90 dB(A) for 8-hour-day exposures. This law was not vigorously enforced and had minimum impact because it applied only to firms with government contracts.

In 1970 the Williams-Steiger Occupational Safety and Health Act was passed. This act applied to all firms engaged in interstate commerce, except for those in the mining and railroad industries. It set up an enforcement organization, the Occupational Safety and Health Agency (OSHA), within the Department of Labor. The National Institute for Occupational Safety and Health (NIOSH) was also established to perform research relating to workplace hazards. Although only their noise-limiting activities are being related here, these organizations are concerned with all types of worker risks.

NIOSH formulated noise criteria recommendations and supplied them to OSHA in August 1972. OSHA then invited comments on these proposed limits, and after reviewing the public response, issued regulations in December 1973 that limited noise levels to 90 dB(A) for 8-hour-day exposures. Like the earlier Walsh-Healy regulations, higher noise levels were allowable if the exposure durations were shorter. For each 5-dB(A) increase in noise, the allowable duration was cut in half. Similarly, if longer exposure times occurred, the level was decreased by the same rate.

Table 1.1 gives the OSHA allowable exposures for a range of workplace noise levels (8). Where worker exposures exceeded these limits, companies are to reduce the levels through engineering or reduce exposures through administrative procedures. The latter course of action meant shifting workers to quieter places for part of a day after they were exposed to high noise levels. This is done in accordance with the rules governing exposures at various noise levels

$$\text{total exposure } C_T = \frac{C_1}{T_1} + \frac{C_2}{T_2} + \cdots + \frac{C_n}{T_n} = <1$$

where

n = number of exposure levels
C_1, C_2, C_n = total time at given level
T_1, T_2, T_n = total time allowed at level

TABLE 1.1 OSHA Allowable Noise Exposures

Noise level [dB(A)]	Exposure per day (hr)
90	8
92	6
95	4
97	3
100	2
102	1.5
105	1
110	0.5
115 max.	0.25

Later, NIOSH recommended a decrease in the 8-hour exposure limit to 85 dB(A). The Environmental Protection Agency (EPA) added its endorsement and recommended that further studies be made of the advisability of further reducing the noise limits. They also objected to using the 5-dB(A) noise-level increment for halving the exposure times; they cite strong technical reasoning for using a 3-dB(A) increment instead. These new proposals were published in the *Federal Register* to solicit public response.

Arguments over noise limits revolve around the fact that workers exposed to high levels become unable to earn a living because their hearing losses make them unemployable. Such hearing losses are therefore eligible for workers' compensation payments, just like those provided for the loss of an arm or leg. It is difficult to say how many workers, subjected to a given noise level, will incur compensable hearing losses, because individuals differ in their susceptibility. Further, some hearing losses are the natural consequence of aging. The scanty available data indicate that 18% of workers subjected to 8-hour-day exposures to 90-dB(A) levels will have compensable losses due to this exposure (9). The percentage incurring losses from exposures to 85 dB(A) levels is 8%. Because these losses occur only after prolonged exposure, it will take a long time to accumulate accurate statistics on the relationship between noise levels and the percentage of workers at risk.

Based on the data that are available, many people feel that too many workers will have their hearing impaired with the present regulations, so they are exerting considerable pressure to have OSHA reduce the limits. This effort was almost successful. Opposition is based on the argument that it will cost industry a prohibitive amount to reduce the levels presently existing in manufacturing plants. At present, as industry is facing stiff foreign competition, this position has prevailed.

It is impossible to predict the future of noise limits. Given the past strong pressure for a reduction and the impetus that more definitive statistics could provide, the limits could be reduced some time in the future. Many machine buyers consider this a real possibility, and since machines last a long time, they demand machine noise levels that still would be acceptable with a reduced ceiling.

1.4 NOISE CONTROL DEMAND CYCLES

One of the peculiarities of machine noise abatement is that the demand for it is cyclical. It has been observed that sometimes people demanding solutions to a critical noise problem have lost interest before solutions were found. Similarly, problems that were ignored for a long time suddenly become very critical. It is as though noise control is a fad that waxes and wanes with time.

Quieting a machine or component takes time. It cannot be accomplished economically without sustained effort. In times when interest becomes low, resources and priorities are withdrawn from the project. When interest is restored, many of the details of the initial work are forgotten and some studies must be repeated.

When it became apparent that the cyclic demand for noise abatement was influencing the effectiveness of noise control programs, an effort was made to determine if they were the result of a company's particular activities or of a more widespread force. Periodical literature indexes were surveyed to see if they provide some measure of the interest in noise. The results of this survey are shown in Figure 1.1.

Two indexes were studied: the "Reader's Guide to Periodical

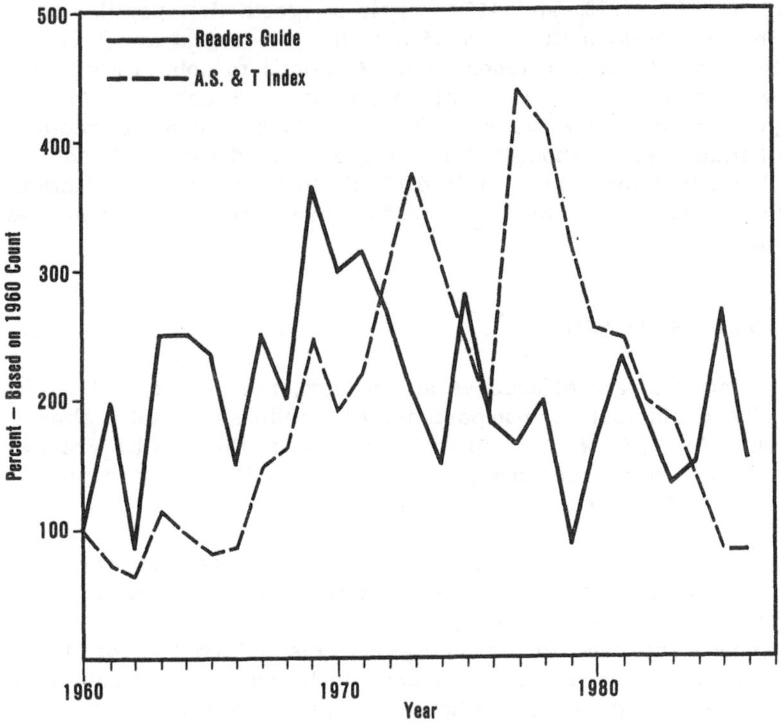

FIGURE 1.1 Noise interest cycles. Magazine noise articles published each year, in percent of number counted for 1960.

Literature" (10), which lists articles appearing in popular magazines such as *Time* and *Reader's Digest*, and the "Applied Science and Technology Index" (11), which lists articles in technical magazines. Articles concerning aircraft, electrical, and undersea noise were not counted. There were many more articles on noise in the technical literature than in the popular press. Some articles were listed under several headings and were counted more than once. It was not practical to ferret out the multiple listings and it is hoped that the degree of error is roughly constant throughout the data. For these reasons the data are plotted in percent, based on the number of articles counted in each index, for 1960. One of the advantages of this procedure is that the scales for the two plots was adjusted to bring them closer together and to make correlations more apparent.

It can be seen from Figure 1.1 that both popular and technical literature output fluctuates widely. Peaks in technical articles appear to lag those of popular magazines. Examples are seen in 1967, 1971, and 1975 peaks in popular literature, followed by peaks in technical literature in 1969, 1973, and 1977. This suggests that popular opinion initiates actions that give rise to later technical articles. If this is the case, there are times where solutions to noise problems are being reported when public interest is at a low ebb, as indicated by the popular literature index. This is analogous to what frequently occurs in industry. Although it is dangerous to draw conclusions from such insubstantial data, it is difficult to ignore the suggestion that both the public and technical sectors have little interest in noise at present.

1.5 GETTING STARTED

The foregoing data are offered as an analogy, not as proof of a theory. They illustrate the importance of timeliness in noise abatement work. All such work requires time, and if the results are to have a maximum commercial impact, the work must be started on the basis of anticipated need and continued without interruption until completed. When the start is delayed until there are strong customer objections, or if effort is allowed to wane when interest appears to slack off, there is a great risk that customers may view even outstanding results with indifference.

Anticipating the need for reducing the noise of existing or proposed machines is not always easy. Data have to be adequate to solicit the support of top management. Although it entails some risks of making wrong judgments, it also has the potential for providing a marketplace advantage that appeals to managers.

1.6 TYPES OF NOISE

Even though we are concerned primarily with audible noise, it is also necessary to consider other acoustic energy forms that can become sound. These are shown in Figure 1.2. Because of frequent concern with these other forms, they are included within the generic term "noise." For this reason we frequently refer to audible noise as *airborne noise*. Vibrations are called *structureborne noise* and fluid pulsations are called *fluidborne noise*. Pumps, for example, commonly generate as much as 1000 times more energy in the form of structureborne or fluidborne noise than they do in the form of airborne noise. These forms act on other machine elements and frequently end up generating more airborne noise than that coming directly from the pump. Because of this, it is important to control all three forms of noise.

Controls for the three types of noise are different. We therefore speak of the forms as separate noises even though they come from the same source. For similar reasons this taxonomy is carried

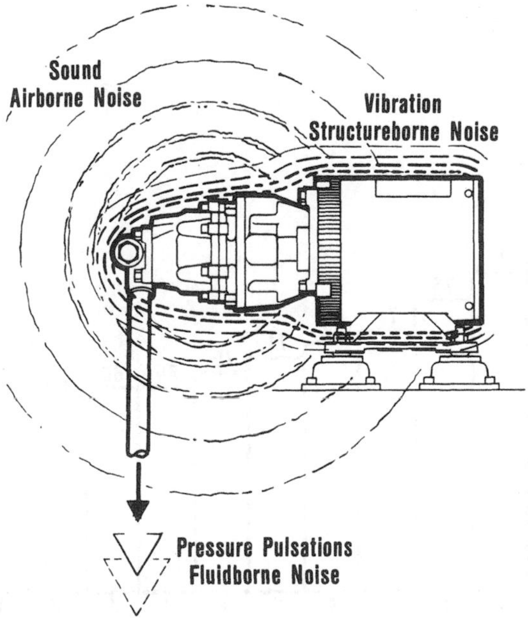

FIGURE 1.2 Three forms of noise must be controlled to achieve low noise levels.

further by referring to, say, structureborne noise traveling through different paths through a machine as separate noises. At times distinctions are also made between different noise frequencies. Even though a machine may have only one noise source, it is said to have many distinct noises.

This is not hair splitting. The control of each of these narrowly defined noises often is unique to that noise. As stated earlier, there are many controls that can be incorporated into a machine that reduce noise. Suppose that a machine makes the noises represented in Figure 1.3 and that we had the means to eliminate any of them, as we wished. Due to the logarithmic nature of the decibel, the total of all of these noises is about 88 dB. Eliminating many of these noises has little effect on the overall noise level of the machine. For example, eliminating noises C through L reduces the total noise by only about 1 dB; it is doubtful if you could hear this difference. Eliminating just noise A reduces the level to about 84 dB, and if noise B is also eliminated, the total drops to about 82 dB. Effective noise control, then, depends on identifying one or two of the strongest noises. Separating noise by energy path and frequency helps identify

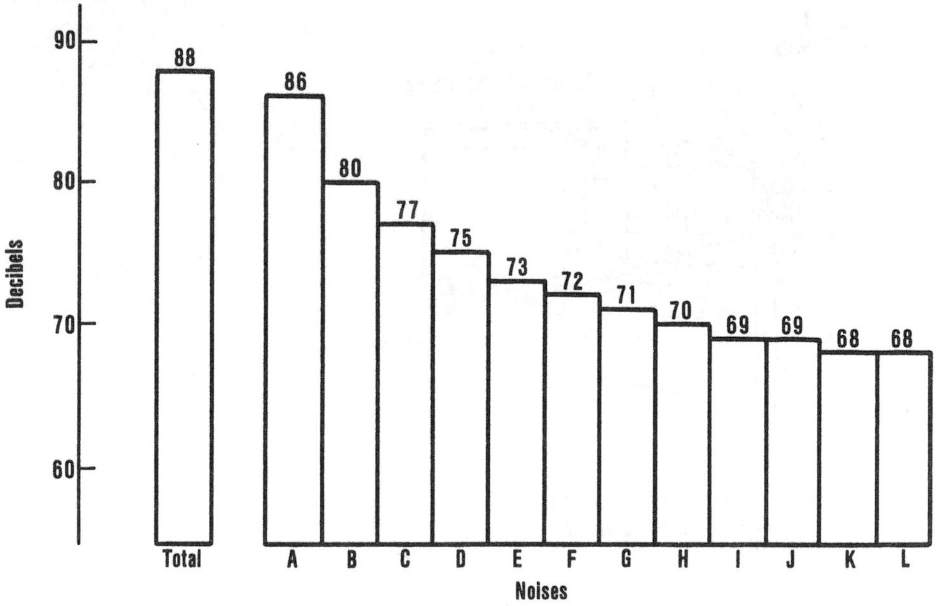

FIGURE 1.3 A machine has many different noises, but most have little effect on the total.

these leading noises. Once we find how they travel from their source to our ear, their control is generally assured.

It would be more accurate to describe these divisions as noise components. This is a bit cumbersome and, it is felt, tends to impede focusing on the critical factors.

1.7 HYDRAULIC NOISE SOURCES

More than 95% of the noise problems in hydraulically actuated machines are traceable to pumps and motors. To date only a few problems have been caused by hydraulic motors because there are relatively few of these in use. However, hydraulic motors are finding greater application in automation machinery, so their use is increasing. They may therefore become a more common source of noise problems in the future.

The infrequency of hydraulic motor noise problems is not the reason for having four chapters devoted to pumps and none to motors. Because of the mechanical similarity between the two devices, most of the pump discussion also applies to motors. Even when motors are mentioned we continue to speak of "pumping chambers" because motors do not have a similar, commonly used term. Information regarding pump port timing will not be directly applicable to motors, however. Here, the reasoning used in designing port timing is explained, so the information can be adapted to motors.

Valves are the other major hydraulic noise source. Most valve noise problems occur in cabs of mobile machines, where they are close to the operator and other noises are screened out. They seldom cause problems in industrial or factory machines, where other noises predominate and tend to mask the valve noise.

1.8 IMPORTANCE OF MEASUREMENTS

Noise measurements are absolutely essential to noise abatement projects. They are needed to determine how much quieting is required and the approach that should be used. Analytical measurements are used to identify the leading noise and how to control it. Finally, they determine the effectiveness of changes and tell when noise abatement goals are met.

This does not mean that noise reduction must be done in an acoustic chamber abundantly equipped with state-of-the-art instruments. A simple hand-held sound-level meter and a relatively quiet work space will be adequate in many cases. This is particularly true in working with machines rather than hydraulic components. Adding sophistication to this minimum recipe improves the accuracy of analyses and usually cuts the cost and time required to reach a given noise goal. Good facilities and instrumentation may be necessary where very quiet machines are required or where a machine has already been subjected to extensive noise control work and additional quieting is needed.

Noise sources in machines are widely distributed and considerable information is derived from measuring how noise levels vary with location. This cannot be done with hydraulic components, which are more compact, so that their noise comes from nearly a point. Because of this, quieting hydraulic components usually requires more than minimum instrumentation to gain the understanding needed for effective quieting. Experience has shown that good facilities and instrumentation are valuable in reducing hydraulic component noise.

1.9 SETTING NOISE GOALS

Setting a noise goal is one of the most important steps that must be taken in initiating a noise reduction project. It provides a basis for estimating the resources and time needed. Once it is set, it serves as a guide, throughout the project, for deciding which proposed steps are taken.

Even when the first noise reduction project is undertaken, goal setting is useful for estimating project dimensions. Without prior experience for guidance, estimates of costs and time are always too low. However, they will be more accurate if goals are set. Further, comparison of the goals and final project costs provides valuable data for planning future noise work.

The guidance derived from noise goals is particularly valuable. It saves time in making decisions. For example, when a machine has levels only 1 or 2 dB higher than a customer's specification limit, only minor design changes are required. In reviewing options, those requiring extensive modifications are given low priority. When a larger noise reduction is needed, only major changes are considered and time is not wasted on changes not having a potential for large reductions.

In setting quieting goals, it is important to gather all applicable information. With an existing product, its measured noise level should be the first data gathered. Noise levels of competitive equipment are very valuable and should be pursued energetically. Market surveys and customer opinions also provide important information. The availability and cost of personnel, facilities, and instrumentation are often overlooked. These factors sometimes have a sobering effect that keeps programs to a practical size.

The proper noise level for a hydraulic component is hard to pinpoint. Actually, it is dictated by the machine in which it will be used and the noise limit for that machine. Machine noise limits are discussed in Chapter 12, and much of the material given is applicable here, for setting component target noise levels.

Some machines attenuate component noise, so they can have units with levels equal to their own allowable limit. Most machines, however, amplify this noise, so they require quieter units. Target levels, then, depend on the types of machines for which the component is marketed. The lower this level, the greater the number of machines for which it will be suitable.

Another way to look at these target noise levels is that when they are low, they require few noise controls when they are used in a machine, so cheaper machines are possible. Because of this, quiet components command a small premium. This, in turn, provides additional incentive for achieving low levels.

REFERENCES

1. B. F. Woodbury, "Noise Can Drive You Crazy!" *Science Digest* 46:1–4, Sept. 1959.
2. R. Brecher and E. Brecher, "Is Noise Getting You Down?" *Saturday Evening Post* 232:32–33, Feb. 6, 1960.
3. J. H. Winchester, "Is Noise Driving You Crazy?" *Science Digest* 56:27–31, Aug. 1964.
4. NFPA T3.9.12, *Method of Measuring Sound Generated by Hydraulic Fluid Power Pumps*, NFPA Recommended Standard, National Fluid Power Association, Inc., Milwaukee, Wis.
5. NFPA T3.9.14, *Method of Measuring Sound Generated by Hydraulic Fluid Power Motors*, NFPA Recommended Standard, National Fluid Power Association, Inc., Milwaukee, Wis.
6. ISO 4412/1, *Hydraulic Fluid Power—Test Code for the Determination of Airborne Noise Levels: Part 1. Pumps*, International Standards Organization, available from American National Standards Institute (ANSI), New York.
7. ISO 4412/2, *Hydraulic Fluid Power—Test Code for the Determination of Airborne Noise Levels: Part 2. Motors*, International Standards Organization, available from American National Standards Institute (ANSI), New York.
8. "Proposed OSHA Ammended Regulations," *Federal Register* 39, Oct. 24, 1974; also "OSHA Criteria on Occupational Noise

Exposure," Title 29, Para. 1910.95, *Code of Federal Regulations*.

9. P. N. Cheremisinoff and P. P. Cheremisinoff, *Industrial Noise Control Handbook*, Ann Arbor Science Publishers, Ann Arbor, Mich., 1977, pp. 5–16.
10. *Reader's Guide to Periodical Literature*, H. W. Wilson Co., New York.
11. *Applied Science and Technology Index*, H. W. Wilson Co., New York.

2
Pumping Forces and Moments

Most hydraulic noise is traceable to periodic variations of internal forces and moments of pumps. It therefore seems appropriate to start the search for ways to reduce noise by examining these sources and their characteristics. In this study we tend to ignore forces that remain constant in magnitude and location or that change slowly, because they do not produce acoustic energy no matter how strong they are.

Inline piston pumps provide a unique opportunity to examine how pumping mechanisms produce noise. The reason is that the pistons move with sinusoidal motion. This permits explicit analysis of motions, pressure profiles, and loads. Such analyses are excellent for illustrating basic pump noise generating mechanisms. For this reason we analyze piston pump forces and moments in detail.

Bent-axis piston pumps do not have true sinusoidal piston motion (1). Their motion is actually quite complex. However, this is an academic consideration. Practically, their piston motion is close enough to sinusoidal to allow them to be treated the same as inline units, in most cases.

Single eccentric vane and external gear pumps are almost as simply analyzed as inline piston pumps, so some detailed analyses of these pumps are also included. Balanced vane pumps have mathematically more complex geometrics and explicit analyses are not feasible. These are best analyzed using numerical techniques and computers, on ad hoc basis, for a particular pump and set of operating conditions. Such analyses do not provide the generalized information that we need here. Therefore, only qualitative analyses of these pumps are made, to highlight their similarities and differences.

There are a great number of different types of positive-displacement hydraulic pumps and it is not possible to analyze all of them here. It is hoped that presenting analyses of the more common pump types, in detail, will make it possible for engineers dealing with other types to adapt the methodology to their pumps as well.

There is concern that it appears that excessive detail is discussed and that some of it seems trivial. Each force and moment has a unique set of characteristics and is carried through the pump structure in a unique way. It is possible to reduce the noise contribution of a given excitation by altering its character or by changing the structure that carries it. Examining these details, then, is expected to suggest options and provide the understanding that leads to effective noise reductions. It also provides the insight needed to interpret noise measurements and other experimental data when searching for the forcing function that is the leading noise contributor.

A number of mathematical analyses are used in this chapter to determine characteristics of various pump forcing functions. These are severely condensed so they do not become tedious and obscure their purpose of illustrating the mechanisms that cause various characteristics. The reader does not have to be a mathematician to benefit from the analyses. In all cases the important results are discussed fully; the mathematics just reinforces the discussion and provides guidance for those who wish to gain additional insight. It is recommended that readers try to duplicate some of the analyses, because it will increase their potential for making noise reductions.

In the analyses in this chapter, angles and angular velocities are expressed in radians. This is done to simplify the analyses. Their use should not cause too much trouble, however. Either degrees or radians may be used in equations in which the angles appear in trigonometric functions. Radians, alone, must be used in the few expressions where angles or angular velocity are one of the parameters. In the discussions, the more familiar degrees are used whenever it is felt that it will not cause confusion.

Most pump airborne noises start as pressure-induced forces. The most effective forces are carried through a pump structure, causing distortions and motions of the outer surfaces and thereby generate sound. These distortions and motions are also the leading source of structureborne noise, as we will see in Chapter 7. The effectiveness of such forces and moments is due to the wide range of their variations. Forces generated by pressure acting on the face of a piston which are carried throughout the pump are a good example of such forces.

Pressure pulsations acting on walls, causing them to act like loudspeaker diaphragms, also generate airborne noise. Although such forces are the leading cause of noise in many hydraulic components, they are generally a secondary source of pump noise.

This is due primarily to the fact that pressure pulsations vary through a range of about 10 to 15% of their maximum values, whereas, as we will see, some forces and moments vary through much greater ranges. Pump structures are designed to withstand maximum force levels, so the ratio of their cyclic range to their maximum is expected to suggest how much the structure responds to the cyclic loading.

Impacts can also be classed as noise-producing forces. In pumps these are often the result of rapid pressure changes. We do not consider them as pressure forces because they produce a distinctly different type of noise. Occasionally, impacts are an important pump noise source, so their characteristics are discussed in some detail in Section 5.4.

Inertia forces from the functional motions of pump parts or imbalance sometimes cause significant structureborne noise. Unless impact is also involved, airborne noise from such forces cannot be important. Structureborne noise from these sources is important only in pumps in some naval applications. Although this is somewhat outside the major thrust of this book, it is discussed because the information may be useful in some research projects.

2.1 PISTON UNITS

Figure 2.1 shows the basic design of an inline piston pump. It consists of a cylinder block that rides on a drive shaft and rotates with it. Pistons, generally an odd number, move back and forth sinusoidally, within cylinder bores in this block, as it rotates. Their motion is due to their contact, through shoes or slippers, with the inclined swash plate. Although pressure presses the pistons and shoes against the swash plate on the pumping stroke, a shoe plate must be provided to maintain the contact during the suction stroke and withdraw the pistons from the cylinder block. In variable-displacement units, the swash plate is mounted on trunnions or similar mechanisms that make it possible to change its inclination and adjust pump displacement or flow.

The cylinder block also bears against a valve plate. Ports in this plate alternately connect the cylinders to the inlet, then to the discharge ports. The switch from inlet to discharge occurs when a piston is at its most extended position. Following engine design practice, we refer to this as bottom dead center. The switch from discharge back to inlet occurs 180° later when the piston reaches its greatest penetration into the cylinder block. This is called the top dead center position.

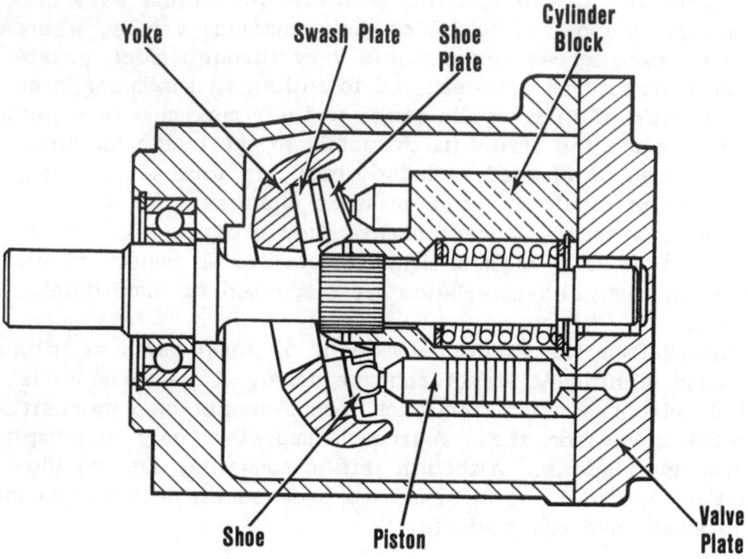

FIGURE 2.1 Inline piston pump.

2.1.1 Single-Piston Forces and Moments

Figure 2.2 shows the geometry and pressure forces of an inline pump. The shaft centerline is the Z axis. The yoke axis, about which the swash plate rotates, is the X axis. The Y axis is at right angles to the other axes and the Y' axis is its projection on the inclined swash plate. The Y axis is called the dead center axis because a piston is at either its bottom or top dead center when its centerline intersects it.

The position of a piston is indicated by the angle θ, measured from the Y axis. The swash plate inclination is indicated by the angle β, which is also measured from the Y axis.

Equilibrium of the piston pressure forces, neglecting friction, is

$$f_z - pA = 0$$

where

- f = reaction force from swash plate, lb
- f_z = axial component of f, lb
- p = cylinder pressure, psi
- A = cylinder cross-sectional area, in.^{2}
- β = swash plate angle, rad

and

$$f_z = f \cos \beta = pA \quad \text{lb}$$

$$f_r = f_z \tan \beta = pA \tan \beta \quad \text{lb}$$

The moment about the X axis (yoke axis)

$$m_x = fr_s \cos \phi \quad \text{in.-lb}$$

where
r = base circle radius of cylinders, in.
r_s = projection of r on swash plate, in.
θ = angular position of cylinder, rad
ϕ = projection of angle θ on swash plate, rad

it can be shown that

$$r_s \cos \phi = \frac{r \cos \theta}{\cos \beta}$$

Therefore

$$m_X = \frac{fr \cos \theta}{\cos \beta} = \frac{rpA}{\cos^2 \beta} \cos \theta \quad \text{in.-lb}$$

Moment about the Y axis (dead center axis):

$$m_Y = f_z r \sin \theta = rpA \sin \theta \quad \text{in.-lb}$$

Moment about the Z axis (shaft torque):

$$m_Z = rpA \tan \beta \sin \theta \quad \text{in.-lb}$$

It should also be noted that there is no force in the direction of the yoke axis, $f_X = 0$.

Both of the forces are functions of the cylinder pressure, p, which only exists during the pumping stroke from bottom dead center to top dead center. It is virtually zero for the suction stroke. For this analysis we assume that the pressure changes are instantaneous, as shown in Figure 2.3. It can be seen that the pressure function is a square wave, having full strength from $0 < \theta > 180°$ and zero for the rest of the cycle. Mathematically, this behavior, for any cylinder, is described by the equation (2)

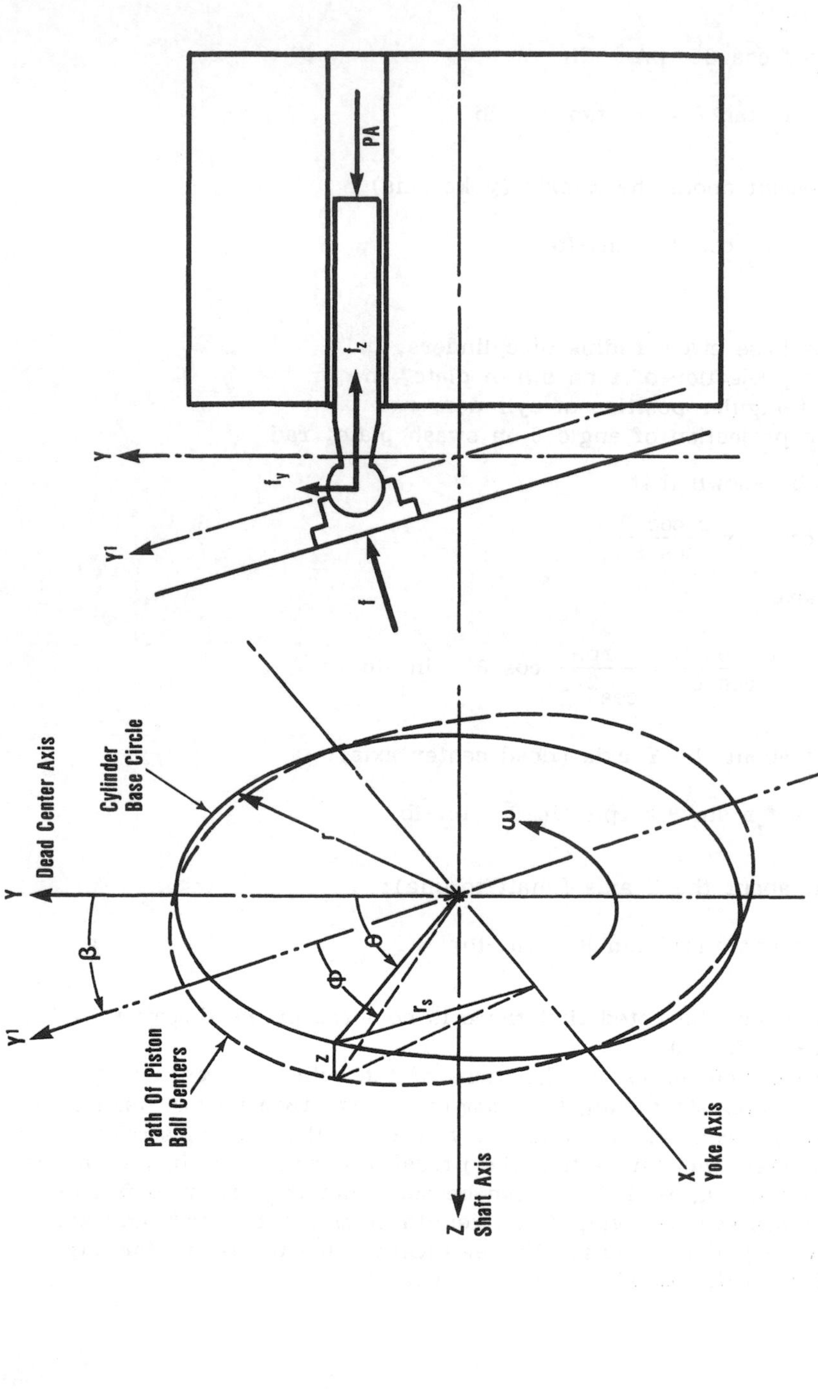

FIGURE 2.2 Inline piston pump geometry.

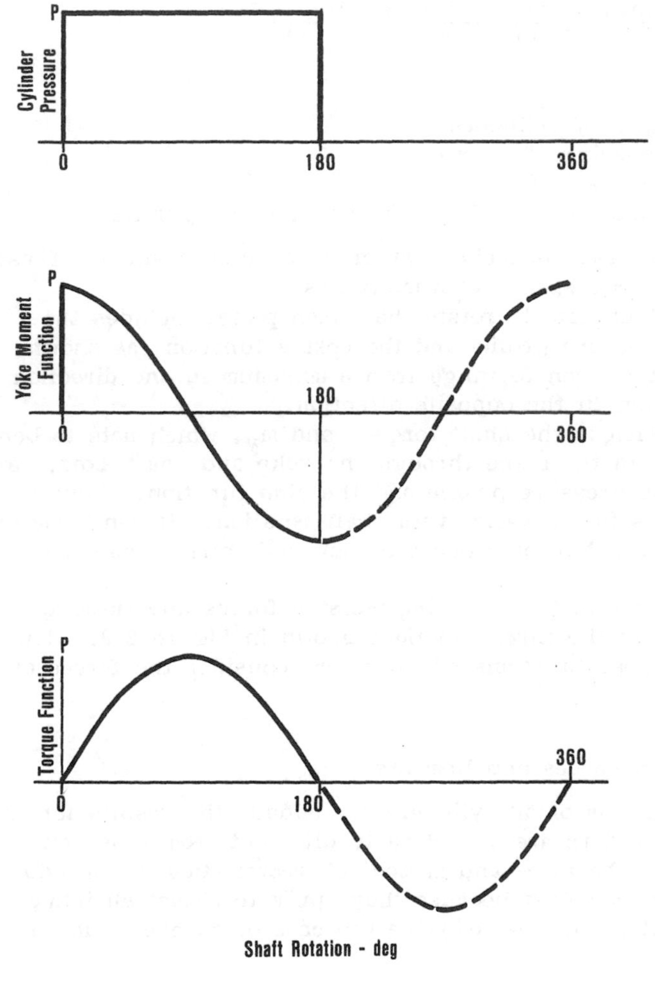

FIGURE 2.3 Inline piston pump: single-piston forcing functions.

$$f(p) = 1 - \frac{|\sin(\theta + k\alpha)| - \sin(\theta + k\alpha)}{2\sin(\theta + k\alpha)}$$

where

n = total number of cylinders
α = cylinder spacing, $2\pi/n$ rad
k = integer, $0 < k < (n - 1)$
$k\alpha$ = angular distance of cylinder from reference cylinder

The moment expressions include either $p \cos \theta$ or $p \sin \theta$. These give the moments some special characteristics.

Moment m_x, which acts to rotate the swash plate, includes the product of this pressure profile and the cosine function, as shown in Figure 2.3. It is seen to range from a maximum in one direction to an equal maximum in the opposite direction.

Moment m_z, which is the shaft torque, and m_y, which acts to bend the pump housing in the plane through the yoke and shaft axes, have the product of the pressure profile and the sine function. Figure 2.3 shows how this factor varies with shaft rotation. It can be seen that these moments act in only one direction and form a half-sine curve.

In summary, we find that the single-piston forces and moments are all described by the three functions shown in Figure 2.3. Now we examine how these functions add when we consider the forces of all pistons.

2.1.2 Total Piston Forces and Moments

When the contributions of all cylinders are added, the results for an odd number of cylinders are considerably different from those for an even number. The force and moment characteristics for an odd number are considered first because they apply to almost all inline piston pumps built. Later we examine the case of an even number of pistons.

Odd Number of Cylinders

If we use the same subscripts as for the single-piston forces and moments, the total forces and moments are

$$F_z = UA \quad \text{and} \quad F_y = UA \tan \beta \quad \text{lb}$$

$$M_x = \frac{VrA}{\cos \beta} \quad \text{in.-lb}$$

$$M_y = WrA \quad \text{and} \quad M_z = WrA \tan \beta \quad \text{in.-lb}$$

where

$$U = \sum_{k=0}^{n-1} f(p)$$

$$V = \sum_{k=0}^{n-1} f(p)\cos(\theta + k\alpha)$$

$$W = \sum_{k=0}^{n-1} f(p)\sin(\theta + k\alpha)$$

Whenever a cylinder reaches bottom dead center ($\theta = 0°$) it becomes pressurized, and the total number of pressurized cylinders is equal to $(n + 1)/2$. Just $180/n$ degrees later a cylinder reaches its top dead center ($\theta = 180°$) and its pressure drops to zero. The number of cylinders under pressure then drops to $(n - 1)/2$. Mathematically, this is expressed as

$$\left(180 - \frac{\alpha}{2}\right) < (\theta + m\alpha) \leqslant 180°$$

where m is the number of pressurized cylinders.

A plot of the total piston force function UA for a nine-piston pump is shown in Figure 2.4. The total piston force consists of a steady force plus a periodic increment equal to the force from one piston. From a noise standpoint the steady component can be neglected. The piston force is then just a square wave with an amplitude equal to the force on one piston and having a frequency of

$$f_p = \frac{n(\text{rpm})}{60} \quad \text{Hz}$$

where

f_p = frequency of piston force square wave, Hz
n = number of pistons
rpm = shaft speed, rev/min

This is called the *piston frequency*.

The summation of the function V, which determines the characteristics of the total yoke moment, is (2)

$$V = \frac{\sin(\theta + m\alpha - \alpha/2) - \sin(\theta - \alpha/2)}{2\sin(\alpha/2)}$$

It is a sawtooth wave, as shown in Figure 2.4. Whenever a cylinder reaches bottom dead center ($\theta = 0°$), this reference cylinder is pressurized and the function achieves its maximum strength. As the

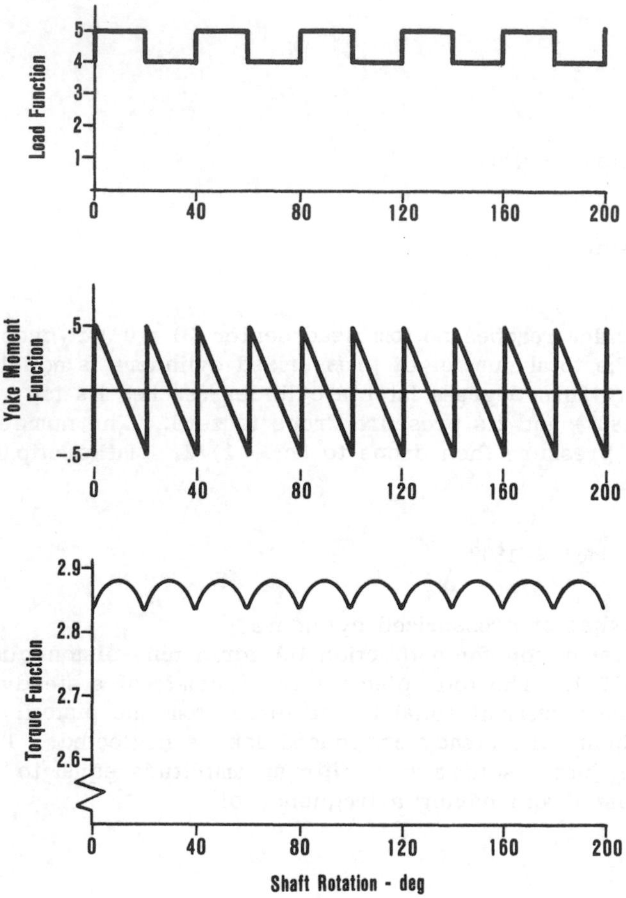

FIGURE 2.4 Forcing functions for a nine-piston pump.

cylinder continues to rotate, the moment decreases linearly, reaching zero when the reference cylinder reaches $\theta = \alpha/4$. The moment continues to decrease linearly until the reference cylinder reaches $\theta = \alpha/2$, where the moment is equal to the maximum moment but negative. At this point another cylinder reaches 180° and is depressurized. This causes the moment to change instantaneously to its positive maximum. Further rotation again causes the moment to decrease linearly until the reference cylinder reaches $\theta = \alpha$. At this point the following cylinder becomes pressurized and the cycle is repeated. Figure 2.5 shows the pressurized cylinder orientations for various stages of the yoke moment cycles. The validity of the function V can be confirmed intuitively by examining this diagram.

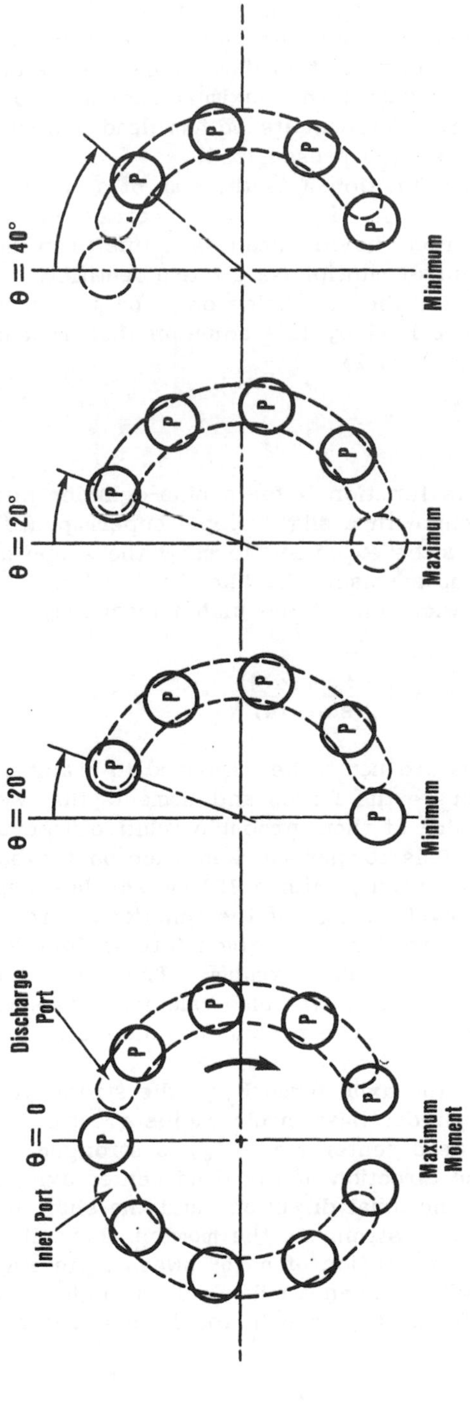

FIGURE 2.5 Pressurized cylinders: nine-piston inline pump.

The yoke moment function completes two cycles in the time interval between two cylinders passing through the same point in their cycles. Its fundamental frequency is therefore twice the piston frequency. It should also be noted that the maximum moment is one-half that produced by a single piston at its bottom dead center if its cylinder was the only one that was pressurized.

The shaft torque function W is the sum of a series of half sine waves like those shown in Figure 2.3. Electrical engineers provide us with this summation, having dealt with this problem years ago in analyzing the output of multiphase electric rectifiers. The following simplified equation for the sum holds only for an interval of $0 < \theta < \alpha/2$. The cycle described by this equation then repeats for each following interval of $\alpha/2$ (2):

$$W = \frac{\cos(\theta - \alpha/4)}{2\sin(\alpha/4)}$$

Figure 2.4 shows function W for a nine-cylinder pump. The result is a constant moment with a slight ripple superimposed on it. Note how the scale had to be expanded to show the shape of this curve. Only the ripple causes noise. It, like the function V, has a fundamental frequency twice that of the piston frequency. The amplitude of the ripple is

$$\text{max. } W - \text{min. } W = \frac{1 - \cos(\alpha/4)}{2\sin(\alpha/4)}$$

Normally, apples are not to be compared to oranges. However, it is desirable to compare the forces and moments that we have discussed so we can get an idea of their probable relative importance in pump noise generation. This comparison was made on the basis of an assumed nine-cylinder pump with a 15° swash plate angle. The maximum level and the cyclic range of the functions were calculated. A structure's stiffness relative to a given forcing function tends to be proportional to the function's maximum. Therefore, the ratio of the cyclic range to the maximum was also calculated as sort of a rough index of each function's noise-producing potential. The results are given in Table 2.1.

In small pumps, the axial force F_z is the strongest forcing function. In pumps with a cylinder base circle radius greater than 2 in., the moment about the dead center axis, M_y, is stronger.

The force in the direction of the dead center axis, F_y, is only about one-fourth that in the axial direction, and the shaft torque M_z is similarly one-fourth as strong as the moment about the dead center axis. This is due to the assumption of a 15° swash plate angle. At smaller angles the ratios will be even smaller. This angle in inline pumps could be as high as 18° (3), which would cause the ratio to be about

TABLE 2.1 Comparison of Pump Forces and Moments[a]

Function	Maximum	Range	Percent
F_z	5pA	pA	20
F_y	1.34pA	0.27pA	20
M_x	0.52pA	1.04rpA	200
M_y	2.83rpA	0.043rpA	1.5
M_z	0.758rpA	0.012rpA	1.5

[a]Nine cylinders, 15° swash angle.

one-third. Yoke angles in bent-axis pumps are as high as 35° (3), which makes the ratio more than two-thirds.

In inline pumps the magnitude of the maximum yoke moment, M_x, is between that of the two strongest and the two weakest forcing functions. In some bent-axis pumps it may be the lowest.

The yoke moment shows by far the greatest potential as a noise source because of its high ratio of range to maximum. The next highest is the axial force, which has only one-tenth the potential. The actual effectiveness of the various forcing functions depends on their relationship to the structure that carries them, and this will be discussed later. This analysis only serves to highlight the two highest potential forcing functions.

Even Number of Cylinders

When a pump has an even number of cylinders the number of pistons under pressure remains constant at all times. When one piston reaches bottom dead center and is pressurized, the cylinder diametrically opposite from it in the block reaches top dead center and is depressurized. Therefore

$$U' = \frac{pn}{2}$$

This constant loading would be ideal because it cannot produce noise. However, pressures do not change instantaneously, as assumed in this analysis. Instead of being constant, the actual force function for an eight-cylinder pump is as shown in Figure 2.6. The effect of this type of pressure profile is similar to that of odd-numbered piston pump loads.

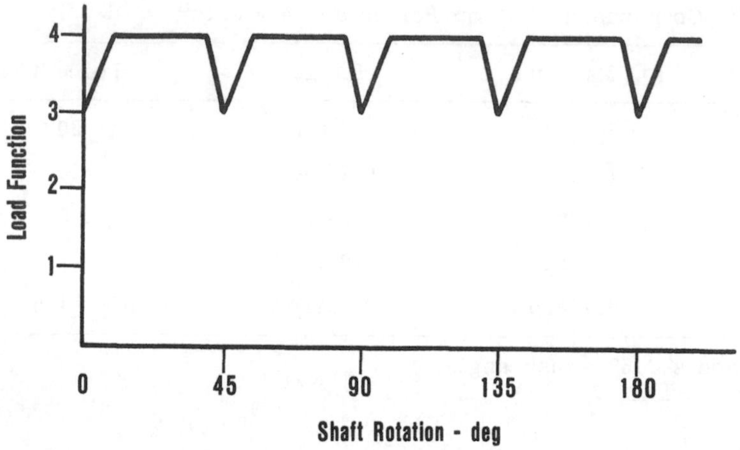

FIGURE 2.6 Effect of pressure rise time in eight-piston inline pump.

The expression for the yoke moment function for pumps with an even number of cylinders is very similar to the one for an odd number. It is (2)

$$V' = \frac{\sin(\theta + 180 - \alpha/2) - \sin(\theta - \alpha/2)}{2\sin(\alpha/2)}$$

Although the expressions are very similar, the actual functions are quite different. The main difference is that the frequency of the yoke moment is equal to, instead of being twice, piston frequency, as in the case of an odd number of cylinders. The function is still a sawtooth and having about the same negative slope. This means that since the period is twice as long, the amplitude is twice as great.

We see a similar difference in the shaft torque function. The expression for this function for an even number of cylinders holds for an interval of $0 < \theta < \alpha$ and repeats for each following interval of $\theta = \alpha$. It is (2)

$$W' = \frac{\cos(\theta - \alpha/2)}{2\sin(\alpha/2)}$$

Again the period is the same as the cylinder spacing and the frequency is the piston frequency. The range of the function is also greater. It can be expressed as

$$\text{max. } W' - \text{min. } W' = \frac{1 - \cos(\alpha/2)}{2\sin(\alpha/2)}$$

It is interesting to compare the torque functions for an odd number of cylinders and one that has twice that number. The angular cylinder spacing for the larger number is only half that of the fewer cylinders. When this fact is factored into the two function expressions, they become identical. If the two pumps have the same displacement and pressure, they have the same torque variations.

2.1.3 Inertia Forces and Moments

The most significant inertia forces in piston pumps are those of the pistons. In Figure 2.2 it is shown that the displacement of the piston is

$$z = r \tan \beta \cos \theta$$
$$= r \tan \beta \cos \omega t$$

where

r = cylinder base circle radius, in.
β = swash plate inclination, rad
θ = cylinder location from dead center
= ωt radians
ω = cylinder angular speed, 2π rpm/60, rad/sec
rpm = cylinder speed, rev/min
t = time, sec

Differentiating the displacement equation provides the piston velocity

$$\dot{z} = -\omega r \tan \beta \sin \omega t$$

The negative sign indicates that the piston is moving into the cylinder when we start measuring. Differentiating again, we obtain the piston acceleration

$$\ddot{z} = -\omega^2 r \tan \beta \cos \omega t$$

This equation indicates that the piston accelerates into the cylinder when we start measuring, t = 0. The swash plate pushes on the piston shoe to provide the force necessary to produce this acceleration

$$f_I = m_p \ddot{z}$$

where

f_I = acceleration force, lb
m_p = piston mass, lb-sec/ft

When the cylinder reaches θ = 90° it begins to accelerate outward. Cylinder pressure provides the necessary force until top dead center is reached and the cylinder is depressurized. The acceleration, therefore the required force, is then at its maximum. The retraction plate, which sandwiches the piston shoes between itself and the swash plate, then provides this force by pulling on the shoe. Up to this point, cylinder pressure clamped the piston, shoe, and swash plate together. Suddenly these three elements are being pulled apart and looseness between them causes impact forces. This is a source of noise discussed in more detail in Section 5.4.2.

Pistons and shoes, for many reasons, are machined to tight tolerances. Their masses, and thus their inertia forces, are essentially equal. Being equally spaced and sinusoidal, the axial forces cancel each other.

The piston centrifugal forces, at right angles to the axial forces, do not cancel because the piston axial locations are staggered. They produce a constant moment that tends to tip the cylinder block. Although this is a major concern to the pump designer, it does not cause noise, so there is no need to analyze it here.

Because of the mass of the cylinder block, if it is even very slightly eccentric to its shaft, a measurable unbalanced force results. This causes a small amount of vibration that can exceed the very strigent structureborne noise specification limits for pumps used on submarines or subnarine tracking ships. However, because the fundamental frequency of this centrifugal force is equal to the shaft frequency, (rev/min)/60, it does not cause airborne noise and is not important in most commercial applications. A rare exception could be a small aircraft pump operating at very high speed.

2.2 VANE UNITS

Vane pumps tend to have somewhat simpler force and moment systems than those of piston pumps. These pumps fall into two classifications: balanced and unbalanced. The unbalanced or single eccentric unit will be considered first because it is easier to analyze.

2.2.1 Single Eccentric Vane Unit

Single eccentric vane units consist of a rotor, with sealing vanes, operating within an eccentrically located ring, as shown in Figure 2.7. The cam surface of the ring is circular and is usually able to move so that its eccentricity, relative to the rotor and shaft axis, is adjustable.

A pumping chamber is the volume trapped between two vanes, the ring and the rotor. It is at its minimum or maximum when it is besected by the axis through the rotor and ring centers. This is analogous to pistons crossing the dead center axis. It would be stretching things too far to apply the engine nomenclature to vane pumps, so we will call these junctures the minimum and maximum points.

We first consider pumps with an odd number of vanes and, therefore, pumping chambers. Because of this, whenever a vane is at the center's axis, the center of a chamber is at the other end of the axis. Again for analysis purposes, it is assumed that a chamber is instantly pressurized when its center crosses the maximum end of the center's axis and is instantly depressurized when its center crosses the minimum end of this axis.

The geometry of the single eccentric is shown in Figure 2.8. To find an expression for the vane motion, we apply the law of cosines

$$R^2 = (r + c)^2 + e^2 - 2e(r - c) \cos \theta$$

where

R = inside radius of ring, in.
r = outside radius of rotor, in.
c = extension of vane from rotor, in.
e = ring eccentricity relative to rotor and shaft center, in.
θ = angle vane makes with maximum side of center's axis, rad

This equation is a quadratic that is solved to provide the equation

$$r + c = e \cos \theta + \sqrt{R^2 - e^2 \sin^2\theta}$$

As long as $c \ll R$, the term $e^2 \sin^2\theta$ is trivial and can be omitted. This simplifies the expression to

$$c = e \cos \theta + (R - r) \quad \text{in.}$$

Since the purpose of this analysis is to illustrate the forcing functions of the pump, it is further simplified by limiting it to the full stroke case of a pump that has $e = R - r$, so

$$c = e(1 + \cos \theta) \quad \text{in.}$$

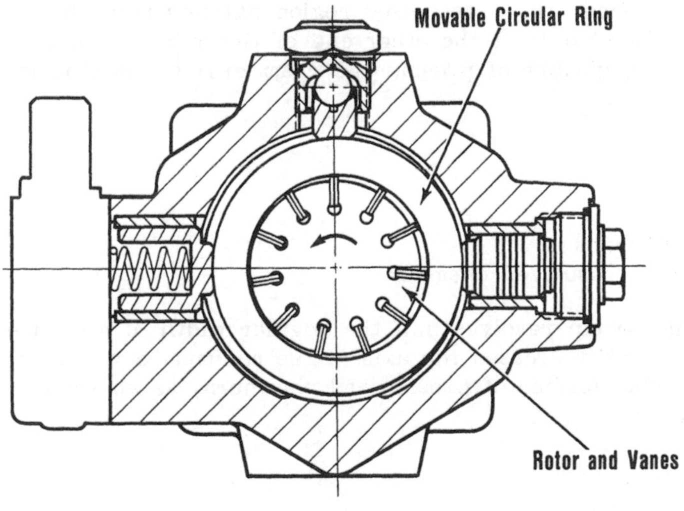

FIGURE 2.7 Variable-displacement single eccentric vane pump.

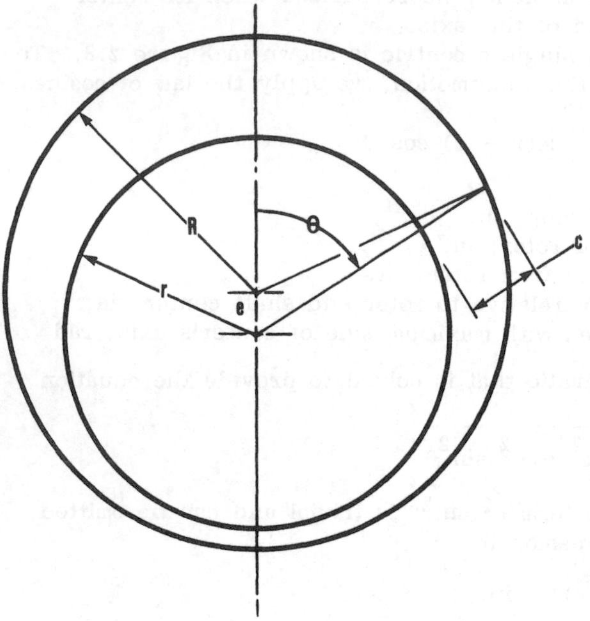

FIGURE 2.8 Single eccentric vane pump geometry.

When a chamber center crosses the maximum end of the center's axis and is pressurized, the pressurized region extends from this segment to the vane that is at the other end of the axis, as shown in Figure 2.9. The number of pressurized chambers at this time is

$$m = \frac{n + 1}{2}$$

where
  n = number of vanes
  m = number of pressurized chambers

After the rotating group revolves half the angular width of a chamber, the center of a chamber crosses the axis at the minimum end and is depressurized. The number of pressurized chambers, as shown in Figure 2.9, is then

$$m = \frac{n - 1}{2}$$

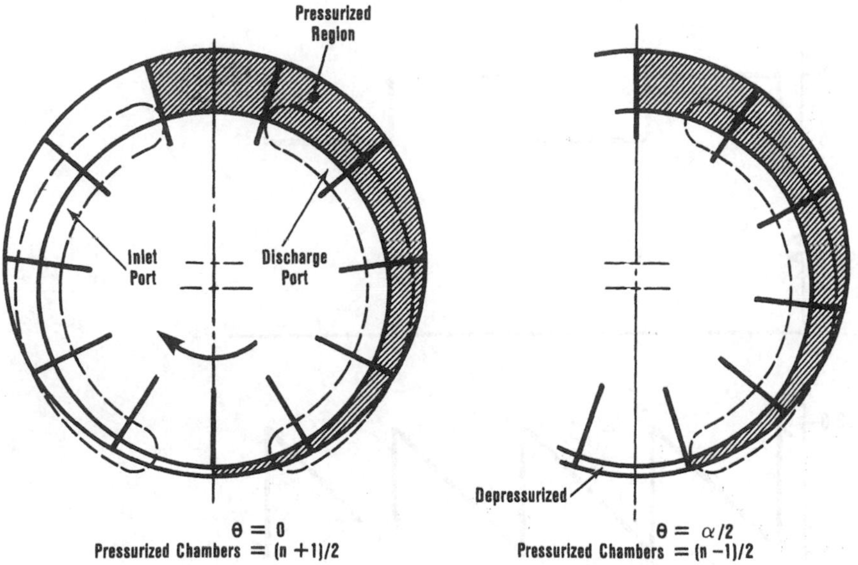

FIGURE 2.9 Pressurized segments: single eccentric vane pump.

This radial force profile is shown in Figure 2.10. It is very similar to the piston pump force profile in Figure 2.4.

The radial load on the ring can be an important source of noise because it must be transmitted from the ring through the housing through the bearings and shaft to the rotor. It is proportional to the number of pressurized chambers

$$F_r = \frac{2\pi Rmwp}{n} \quad \text{lb}$$

where

p = discharge pressure, psi
w = width of ring, in.

Only a portion of this load reaches the bearing, however. Because of the circular geometry of the ring and housing, some of the load is canceled within the structure. This mechanism is discussed in more detail in Section 2.3.5. The portion reaching the bearing is equal to the product of the pressure and the area of a chordal plane through the rotor that spans the pressurized chambers, as shown in Figure 2.11.

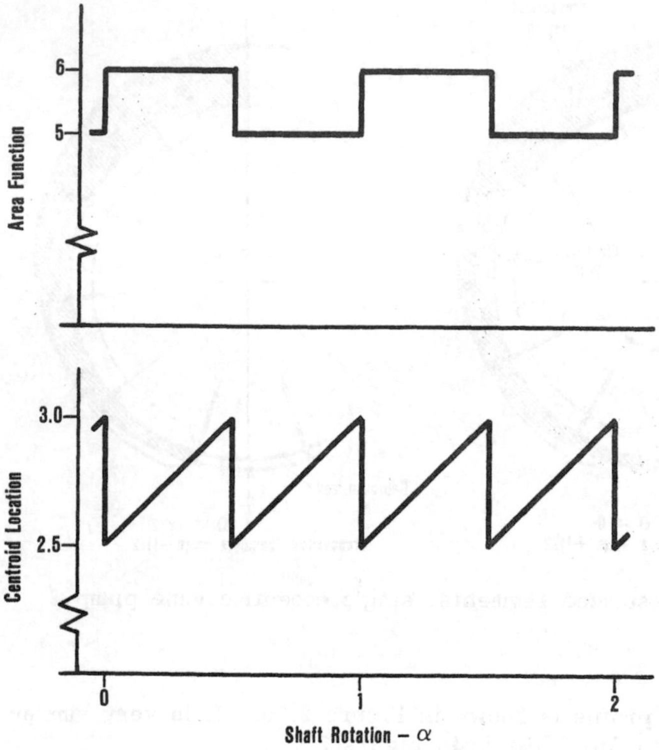

FIGURE 2.10 Radial load: single eccentric 11-vane pump.

$$F_b = 2pwR \sin \frac{\pi m}{n} \quad \text{lb}$$

We are most interested in the change in the radial force on the ring. This change is equal to the force from one pumping chamber

$$\Delta F = \frac{2\pi Rwp}{n} \quad \text{lb}$$

Because of the sine term in the bearing load equation, its increment is not as easily determined. It must be found by taking the difference in the forces calculated with the previous bearing load equation, while using $m = (n + 1)/2$ and $m = (n - 1)/2$.

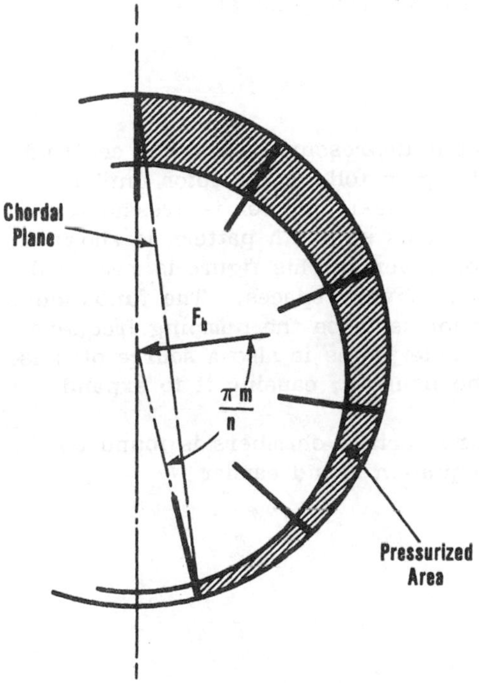

FIGURE 2.11 Bearing load: single eccentric vane pump.

Not only the magnitude but also the location of the loads changes. Instead of finding how this affects the various moments, as in the piston pumps, we examine how the centroid shifts position. The centroids of both the radial and bearing loads lie on the same rotor radius, so we analyze only the centroid of the radial load.

When a chamber is pressurized, the centroid is located at its starting point at

$$\gamma = \frac{\pi}{2} - \frac{\alpha}{4} \quad \text{rad}$$

where

γ = angular location of radial force centroid, rad
α = angular spacing of vanes, $2\pi/n$ rad

The centroid rotates with the rotor until the center of the leading pressurized chamber reaches the minimum end of the center's axis.

The centroid is then located at its farthest point at

$$\gamma = \frac{\pi}{2} + \frac{\alpha}{4} \qquad \text{rad}$$

At this instant the leading chamber depressurizes and the centroid returns to its starting point. It again follows the rotor until it reaches its farthest point, when the next chamber is pressurized and it returns to the starting point. This sawtooth pattern is shown in Figure 2.10; the centroid location given in this figure is measured from the max center axis in terms of vane spaces. The fundamental frequency of this radial load motion is twice the pumping frequency.

Pressure acting on the pump side plates is also a source of noise. This force is carried through the housing, causing it to expand axially with periodic motion.

The transverse area of the pressurized chambers is found by integrating the vane extension equation found earlier

$$A_t = \int c\left(r + \frac{c}{2}\right) d\theta$$

The result of this process is

$$A_t = e\left[\left(r + \frac{3e}{4}\right)\theta + (r + e)\ \sin\ \theta + \frac{e}{8}\ \sin\ 2\theta\right]_{\theta=\theta_1}^{\theta_2}$$

Calculations for this area were made with this equation, for a pump with the following parameters:

$R = 1.6$ in. $\quad n = 11$
$r = 1.5$ in. $\quad e = 0.1$ in.

The resulting transverse area or axial force function is shown in Figure 2.12 for both full and zero stroke.

This force increases sharply when a new pumping chamber is pressurized. It then decreases with rotation for an interval equal to one-half a chamber width. The rate of decrease is greatest at full stroke and zero at zero stroke, assuming that there is pressure at zero stroke. An abrupt drop in load occurs when a chamber becomes depressurized. At full stroke the area of the chamber at this point, in the example pump, is so small that the drop does not show in our plot. The load continues to decrease for another increment of rotation equal to half a chamber width and then the cycle repeats.

It can be seen that the exact shape of the curve depends on the stroke adjustment and the minimum chamber volume at full stroke.

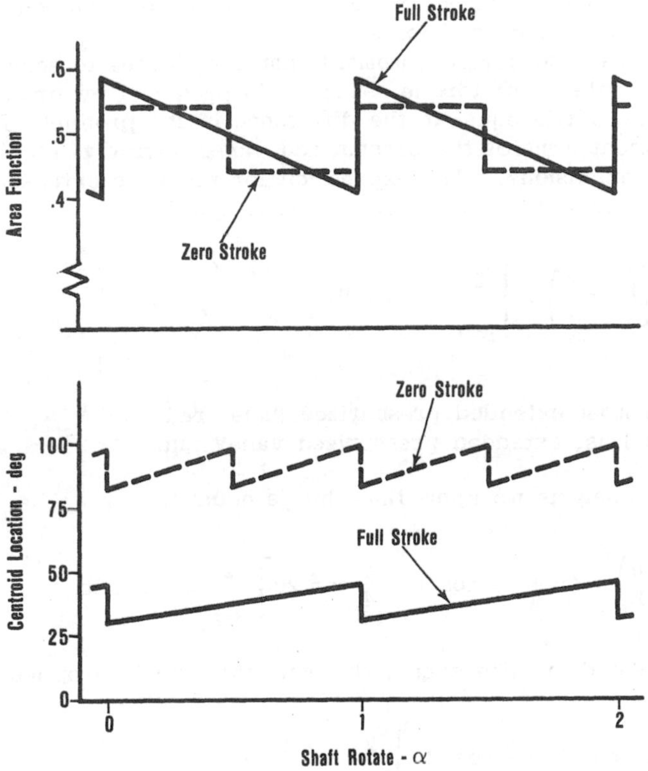

FIGURE 2.12 Axial load: single eccentric 11-vane pump.

At small stroke settings the load function is primarily a square wave at pumping frequency. As full stroke is approached, the function becomes a sawtooth with the same frequency.

As in the case of the radial load, the position of the axial force changes and this, too, causes sound. Unlike the radial loads the location of the axial load centroid cannot be determined by inspection, except at zero stroke. At this condition, all chambers have the same transverse area, so the axial load centroid follows the pattern of the radial load centroid.

The full-stroke motion of this centroid was calculated from a derivative of the vane extension equation and is shown in Figure 2.12 together with that of the zero stroke. This was done to complete the illustration but is not worth the trouble in a practical case. For that reason the details of the calculation are not given here. It can be seen that the frequency characteristics of the axial load centroid are the same as those of the axial force.

The vane pump has one pumping moment that contributes to sound generation. This is the shaft torque. Torque is produced by pressure acting on the vanes. It is equal to the difference in the products of the forces and moment arms of the pressurized vanes having the greatest and least extensions. The expression for the vane extension is utilized in determining these products. The torque is

$$\text{torque } T = pw\left[\left(r + \frac{c}{2}\right) c\right]_{\theta=\theta_1}^{\theta_2} \quad \text{in.-lb}$$

where

θ_2 = location of most extended pressurized vane, rad
θ_1 = location of least extended pressurized vane, rad

From the previous analysis we know that this is equal to

$$T = pwe\left[\left(r + \frac{3e}{4}\right) + (r + e) \cos \theta + \frac{e}{4} \cos 2\theta\right]_{\theta=\theta_1}^{\theta_2} \quad \text{in.-lb}$$

Since we are interested in differences, the constant term is dropped

$$T = pwe\left[(r + e) \cos \theta + \frac{e}{4} \cos 2\theta\right]_{\theta=\theta_1}^{\theta_2} \quad \text{in.-lb}$$

For our example pump this becomes

$$T = pw\left[0.16 \cos \theta + 0.0025 \cos 2\theta\right]_{\theta=\theta_1}^{\theta_2} \quad \text{in.-lb}$$

The torque function for the example pump, at full stroke, is shown in Figure 2.13. The range is so small that a greatly expended scale had to be used. It is seen that the shape of this function is triangular.

As in the case of the piston pump, we calculated the strength and cyclic range of the forces and the torque, and their ratio, to gauge their noise production potential. The pump parameters used are the same as those given earlier for use in calculating the forcing functions. The results are given in Table 2.2.

The axial force has the highest potential for noise production, although the radial force levels are higher. Both are strong and

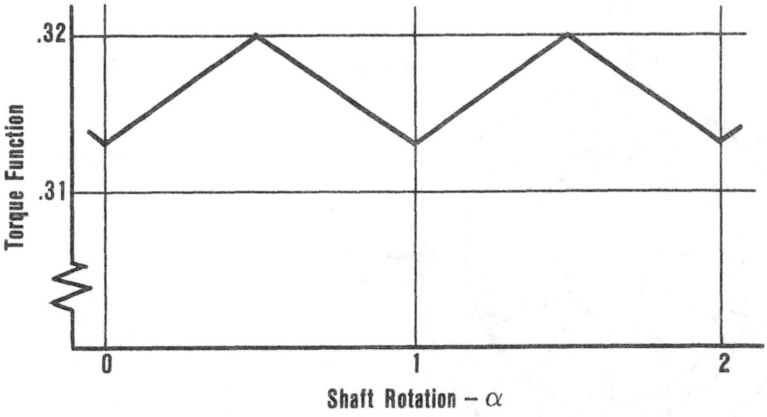

FIGURE 2.13 Torque: single eccentric 11-vane pump.

are expected to be significant noise sources. The torque, on the other hand, although strong, has a small cyclic range and is not thought to be a leading noise source.

The characteristics of the forcing functions of pumps with an even number of vanes are easily predicted from these analyses. By this process it is seen that such pumps would have no radial force variations, while transverse force and the torque variations would be about the same as those for a pump with one more or less vane. These observations, of course, are based on the assumption of instantaneous changes in pressure in the pumping chambers.

2.2.2 Balanced Vane Pumps

Balanced vane pumps are so named because their radial loads are balanced within their cam ring. This eliminates the high bearing

TABLE 2.2 Comparison of Vane Pump Forces and Moments[a]

Function	Maximum	Range	Percent
Radial	2.74wp	0.457wp	17
Axial	0.586p	0.181p	31
Torque	0.320wp	0.007wp	2.2

[a]Eleven vanes, 0.1 in. eccentricity.

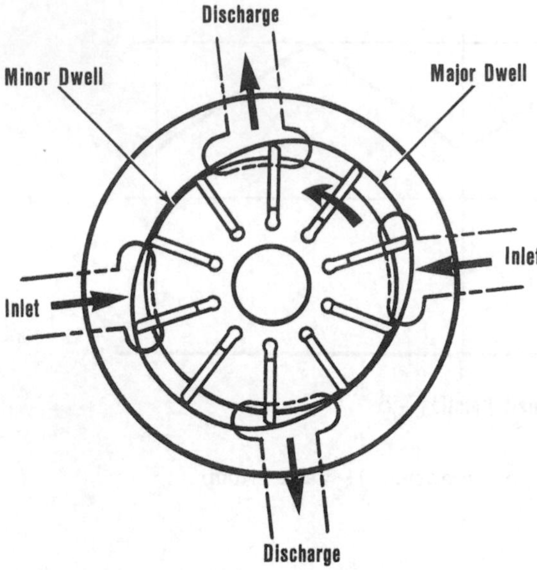

FIGURE 2.14 Balanced vane pump.

loadings that occur in single eccentric vane pumps. What is most significant from a noise standpoint is that most of the radial load is carried internally with minimal effect on the outer radiating surfaces.

Load balancing is accomplished by providing a cam ring, concentric with the shaft and rotor, that has two identical cam profiles positioned diametrically opposite each other, as shown in Figure 2.14. All of these pumps have an even number of rotor vanes. As a pumping chamber goes through its cycle, there is always a diametrically opposite chamber at the same pressure.

The complete pumping cycle has to occur in half of a shaft revolution, so cam design is critical. Figure 2.15 shows the basic elements of these cams. Constant-radius cam segments, called dwells, are provided wherever commutation between the inlet and discharge ports has to occur. These preempt a large segment of the cam ring and reduce the time available for the vanes to move in and out of the rotor. Since the dwells have to be equal or greater than the spacing between vanes, the tendency is to use as many vanes as possible so that the dwells will be small and there will be more time for vane motion.

For the last 50 or more years cam designs have been under continuous development to increase speed capability and displacement

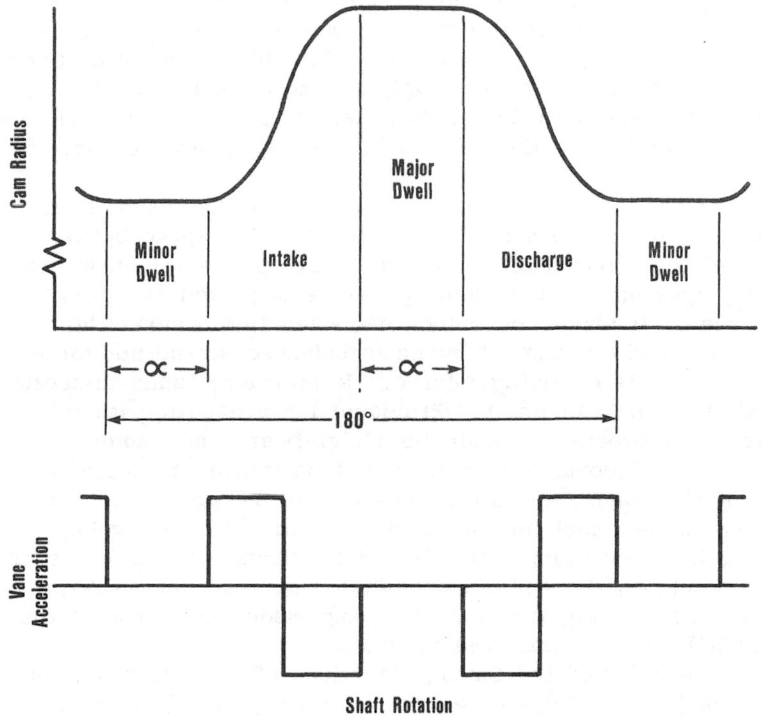

FIGURE 2.15 Balanced vane pump cam design.

for a given frame size. Fortunately for our purposes, we do not have to deal with the refinements in these cams.

Basically, each cam consists of a rise and a fall. These, in turn, are divided into two equal sections that provide constant acceleration to the vanes, as shown in Figure 2.15. The first accelerates the vanes from their zero-velocity state in a dwell and the second decelerates them until they again have zero velocity at the start of the next dwell. The outward acceleration, which occurs in the suction part of the cam, determines the extension that the vane can achieve. It is limited by the available forces. Vane centrifugal force assists but is not sufficient to achieve the extensions needed for modern pumps. The rest of the needed force is supplied by discharge pressure acting on the underside of the vane. This pressure is also needed to counteract the pressure acting on the vane tips in the high-pressure regions of the cam.

If discharge pressure is applied to the entire underside of the vane, excess wear occurs in the suction section of the cam, where only low pressure acts on the vane tips. For this reason vane pumps generally have some provision for applying the pressure to only a part of the undervane area during this part of the pumping cycle and over the entire area when the vane is in the high-pressure part of the cycle.

Figure 2.16 shows an intervane that is used to do this in some pumps. Discharge pressure is always ported to the space between the vanes and the intervanes. When the vane is in a region where pressure impinges on its outer end, pressure is ported to the entire undervane area. It should be noted that when this occurs, the pressure on the ends of the intervane is balanced so the net force on this element is its centrifugal force. It therefore tends to accelerate toward the vane and hit it. Provision for preventing impact must be provided because it would be a significant noise source.

Since a simple algebraic equation cannot be written to describe the motion of the vanes, an explicit analysis is not possible. Computer numerical techniques must be used and calculation of the forcing functions for balanced vane pumps is quite tedious. Results of such calculations would not be general enough for our present purpose, so we evaluate these forcing functions by comparison with those of the pumps, which have been analyzed in detail.

The characteristics of the forcing functions of this pump are very similar to those of the single eccentric vane pump at full stroke. When the number of vanes in a balanced vane pump is even but not divisible by 4, its forcing functions will be similar to those of an

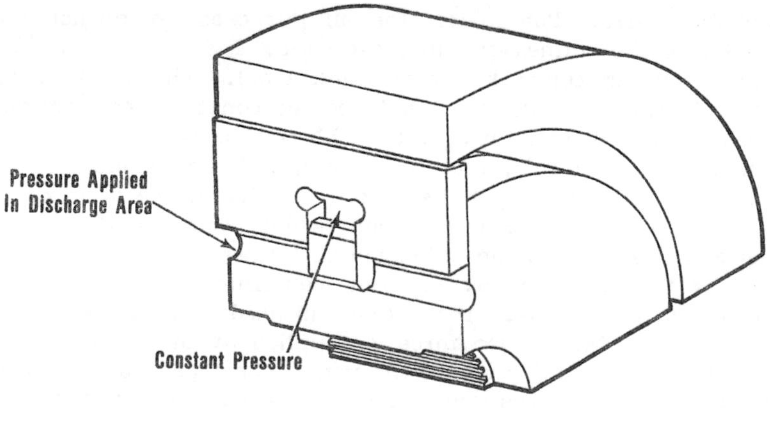

FIGURE 2.16 Intervane.

unbalanced vane pump with an odd number of vanes. As discussed earlier, the radial forces at the two cams cancel each other within the cam ring. The axial forces acting against the side plates, however, add together in elongating the pump. Torque fluctuations also add together. No generalizations can be made concerning the amplitudes of the forcing functions since they depend on the characteristics of the cams, whose shape is indeterminate.

A balanced vane pump whose number of vanes is divisible by 4 has forcing functions similar to an unbalanced vane pump with an even number of vanes. With instantaneous pressure changes, the radial forces are constant and the tangential forces and torque are similar to those of an unbalanced vane pump with an odd number of vanes.

Although the author knows of a very quiet 12-vane pump, it is believed that this is due only partially to the factors being discussed here. Probably much is due to improved pressure control, which is a subject of Chapter 5.

2.3 GEAR PUMPS

Gear pumps use the spaces between gear teeth as pumping chambers. When these chambers move from the inlet to the discharge port, they are sealed by a circularly contoured stationary member that is closely fitted to the tips of the teeth. The chambers return to the inlet through a mesh of two gears which seals between the inlet and discharge ports.

A great number of different conjugate rotor shapes are used in gear pumps; however, this discussion is confined to those with involute teeth, which are the most common. Analyses of forcing functions are made for external gear pumps, such as those shown in Figure 2.17, and it is expected that the reader can adapt these to internal gear pumps like the one shown in Figure 2.18.

Gear pump noise is not produced by the same mechanism that produces noise in gear drives. The latter noise is due primarily to shock occurring as teeth come in contact. When two gear teeth mesh, they remain in contact as their following teeth come into contact. If the geometry of the gears is perfect, the incoming teeth slide into mesh smoothly. However, tooth deflections caused by the transmitted loads and machining errors misalign the incoming teeth and they collide as they engage.

Pump gears must be very accurately made to provide good hydraulic performance. Since they are generally of a coarse pitch (4), they are relatively stiff. In addition, they are designed to have two pairs of teeth in contact most of the time. This means that they never carry their entire load near the tips of their teeth, where it would

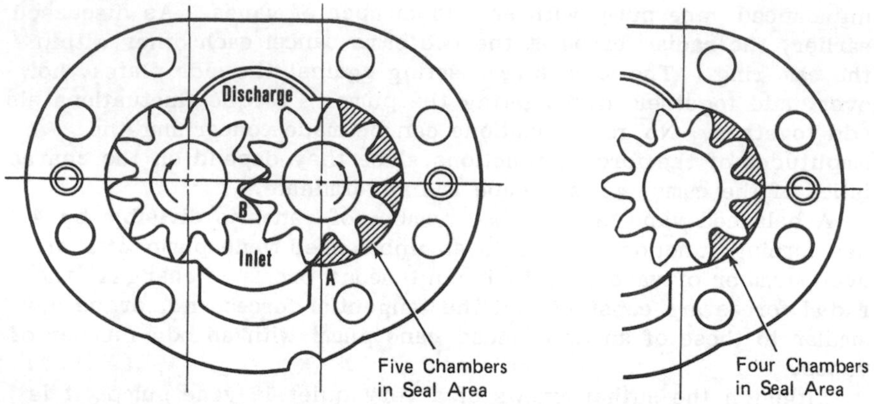

FIGURE 2.17 External gear pump.

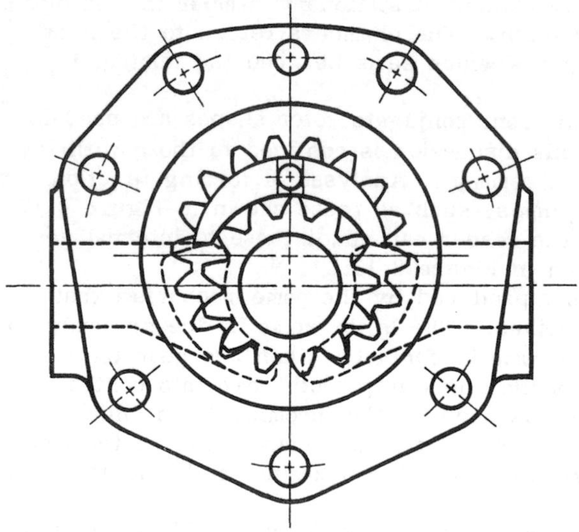

FIGURE 2.18 Internal gear pump.

produce the maximum deflection. Both of these factors minimize tooth contact shock. Further, the pitch line velocities, which are another important gear noise factor, are not high. Because of this it is believed that pumping pressure forces, not contact shock, are responsible for gear pump noise. It must be noted, however, that this cannot be verified by analyzing the noise because both mechanisms produce noise at the pumping frequency and its harmonics.

2.3.1 Gear Tooth Mechanics

Although gear operation is very simple, few people are acquainted with its details to the extent needed to analyze gear pump forces. A short review of gear mechanics is therefore in order. In particular, we need to find the location of the various tooth contact events. This requires the use of involute geometry, which if we wish to make it fully understood would require an extensive tutorial. This would make this discussion very tedious, so a briefer treatment is given. Those wishing for a better understanding will find Buckingham's classic text (5) very interesting, as well as helpful.

The involute shape of gear teeth is generated by the end of a string as it unwinds from a circle, called the *base circle*. The base circles of two gears that are to mate can be at any distance from each other and the curves generated from them will transmit uniform motion between them. The speeds of the two gears is in proportion to their base circle radii. This flexibility in selecting center distances is the unique advantage of involutes.

A tangent to the base circles, crossing the centerline, establishes the *pitch point* at the intersection. This point is P in Figure 2.19. Circles about the gear centers that pass through the pitch point are called pitch circles.

The pitch point divides the distance between the two centers in proportion to the base circle radii of the gears. In our case of identical gears it divides the line in half. When the point of contact between gear teeth is at this point, the two segments of the tangent coincide with the lines that generated the two mating surfaces.

The angle between the centerline and the radius to the point of tangency, b, is the pressure angle, ϕ. This is the parameter that is generally used to describe the relationship between the base circle radii and the distance between their centers. It also determines the proportions of the gear tooth profiles.

The tangent line is called the *action line* because the locus for all the tooth contacts are along this line. It is well to note that the angle between the action line and a perpendicular to the centerline is also the pressure angle.

There is one other parameter that determines the proportions of gear teeth. It is the *circular pitch*, which is the circumference of

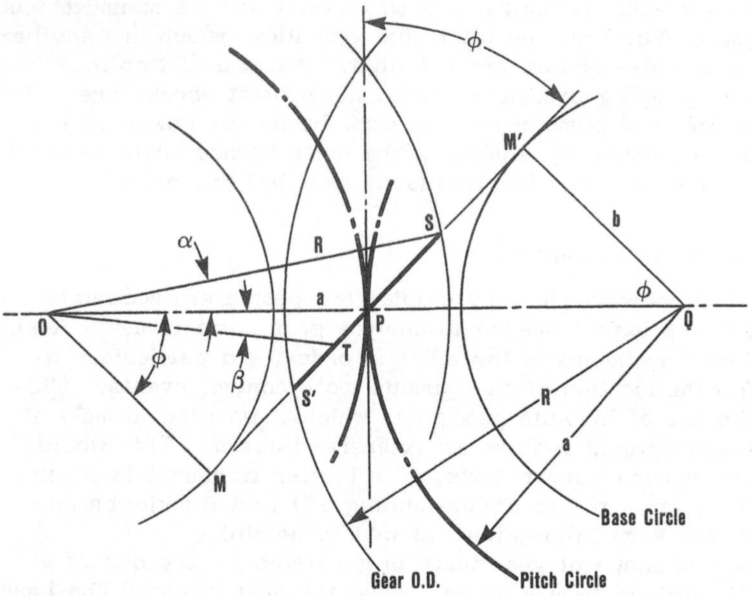

FIGURE 2.19 External gear pump geometry.

the pitch circle divided by the number of teeth. Mating gears always have the same circular pitch.

There is also a *normal pitch*, which is equal to the circumference of the base circle divided by the number of teeth. It is the spacing between teeth as measured along any tangent to a base circle. Returning to the concept of the end of a string generating a tooth profile, if the string around the base circle circumference is marked off at intervals equal to the number of teeth, each of the marks would generate the involute face of a tooth, as shown in Figure 2.20. Since the base circle circumference was divided by the number of teeth, the spacing of the teeth, as measured along a tangent to a base circle, is equal to the normal pitch. Since the action line is one of these tangents, we know that contacts that travel along this line are spaced one base circle pitch apart.

Contact between two teeth starts when the tip of the driven gear tooth reaches the action line, where it contacts the base of a tooth of the drive gear. Contact continues along this line until the tip of the drive gear tooth reaches the line and loses contact with the base of its driven tooth. Before this occurs, the trailing pair of teeth are in

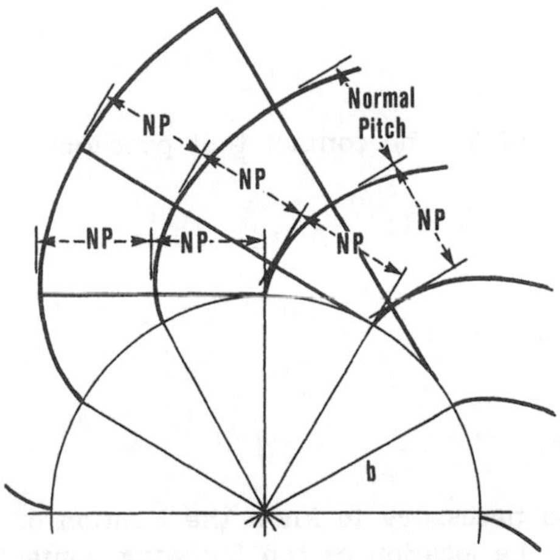

FIGURE 2.20 Generating gear tooth involute profiles.

contact so that the seal between the discharge and inlet ports and continuity in the motion transfer is maintained.

In Figure 2.19 the distance along the action line between the pitch point and contact initiation is

$$SP = SM - PM \quad \text{in.}$$

where

$SM = \sqrt{R^2 - b^2}$
$PM = a \sin \phi$
a = pitch radius of gears, in.
b = base circle radius = $a \cos \phi$, in.
R = outside radius, in.
ϕ = pressure angle, rad
n = number of teeth
θ = gear rotation angle measured from centerline, rad

The angluar position of the first contact point relative to the centerline is

$$\alpha = \sin^{-1} \frac{SM}{R} - \phi \quad \text{rad}$$

Similarly, tooth contacts travel along the action line in direct proportion to gear rotation. Displacement along the line measured from the pitch point is

$$SP = b\theta$$
$$= \theta a \cos \phi \quad \text{in.}$$

We also need to know the location of T, the contact that preceded the one being initiated

$$PT = SP - \frac{2\pi b}{n} \quad \text{in.}$$

The angular position of point T is

$$\beta = \phi - \tan^{-1} \frac{PM - PT}{b} \quad \text{in.}$$

In the force analyses it is also necessary to know the location of the farthest point of contact and the location of the following contact point when contact between two teeth is ended. Because the two gears are identical, the geometry is symmetrical about the pitch point. The desired points are therefore mirror images of the ones that have already been located.

If gears are made with the theoretically correct thickness, there would be contact on both sides of the teeth, as seen in Figure 2.21. This encloses two fluid volumes, at A and B. The one at A is moving toward the pitch point, where its volume will be its smallest. The volume is therefore decreasing and compressing the trapped fluid. Very high pressures are generated in this way. Similarly, the pocket at B is expanding and reduces its pressure to where cavitation occurs.

Both of these conditions are extremely undesirable. The gears are thus cut slightly thinner than the theoretically correct thickness, as shown in Figure 2.21. This allows communication between the two trapped volumes, so their combined volume will tend to remain constant. It can be seen that this does not affect the seal between the ports.

Although providing clearance is both necessary and effective, it is not adequate. Additional relief of the trapped pressure is generally provided by pockets in the side plates, like that at A in Figure 2.22, which communicate with the discharge port. The fluid between gear contact points is therefore essentially at discharge pressure.

The test for the adequacy of these measures is operation at very low discharge pressure. If there is still a pressure buildup in the trapped tooth spaces, the increased pressure will cause the gears to rattle audibly.

2.3.2 Discharge Port Forces

The pumping chambers between gear teeth are subject to pressure from the time they close to the inlet port at A in Figure 2.17 until they complete their passage through the gear mesh at B. Unlike

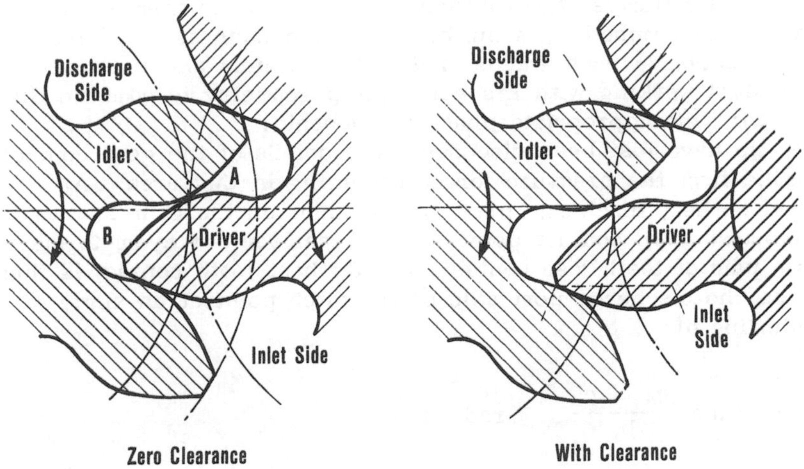

FIGURE 2.21 Pump gears must have clearance.

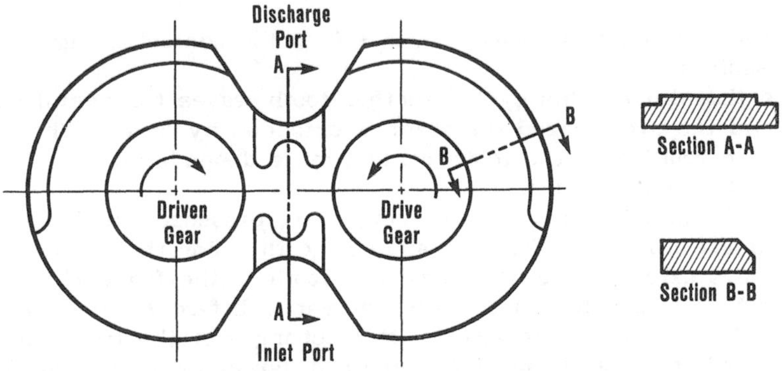

FIGURE 2.22 Gear pump side plate.

piston and vane pumps, the rotation distance between them is not fixed and is determined by a number of design factors. For this reason the action at each end is analyzed separately and it is left to the analyst working with a specific pump geometry to combine the analyses with the appropriate phase relationship.

Since we have just been discussing the details of the gear contacts we examine first the pressure forces occurring in that area. As mentioned earlier, discharge pressure impinges on the gear surfaces up to the last tooth contact point. If we consider our cycle to start just as a pair of teeth lose contact, the farthest tooth contact is then located on the discharge port side of the pitch point at an angle with the centerline of

$$\beta = \phi - \tan\frac{PM - PT}{b} \qquad \text{rad}$$

The contact point then travels to the point where contact ends, on the inlet port side of the pitch point, at an angle to the centerline of

$$\alpha = \sin^{-1}\frac{SM}{R} - \phi \qquad \text{rad}$$

During this time the gears rotate through an angle of $2\pi/n$. The pressurized zone therefore decreases by

$$\Delta Z = \alpha + \beta - \frac{2\pi}{n} \qquad \text{rad}$$

The zone then suddenly decreases by $\alpha + \beta$ as the contact being tracked dissolves.

At some time during this cycle, another tooth leaves the cylindrical seal area and the pressurized quadrant is extended by $2\pi/n$. Where this occurs in relation to the tooth contact cycle depends on the geometry of the specific pump.

It should be noted that on the driving gear, pressure extends from the front of a tooth to the front of another tooth. For the driven gear, pressure extends from the back of a tooth to the front of another tooth. It therefore has pressure over a different arc than that of the driving gear. Because of this, changes at the two ends of the pressurized zone of one gear have a different phase relationship than that of the other gear.

This changing pressurized zone is felt directly by the bearings, together with pressure acting in the seal area. This is analyzed in a later section.

Some of this changing discharge port zone also contributes to the housing load. This contribution is not affected by the gear mesh geometry. It provides a sawtoothed load-time variation caused by teeth leaving the cylindrical seal surface and opening the area to the following tooth.

For analyzing axial force from the discharge port area, we neglect the tooth mesh area because it has very little transverse area. For this purpose, then, the discharge port area is taken to be bounded by the first rather than the last contact point. The other boundaries, teeth leaving the circular seal area, are the same as for analyzing the radial loads.

The total transverse area of this region depends on the design and cannot be determined here. We therefore examine only the changes in this area. To this end we replace the gears with vanes extending from the gear centers to the seal points, as shown in Figure 2.23.

The instantaneous change in the discharge port volume is equal to the difference in the volumes being swept by the replacement vanes

$$dV = \frac{w}{2}\left[(OR^2 - OS^2) + (QT^2 - QS^2)\right] d\theta \quad \text{in.}^3$$

where

V = volume of discharge port sector, in.^3
θ = gear rotational angle, rad
w = gear width, in.
R = gear outside radius, in.
= OR = QT
a = gear pitch radius, in.
= OP = PQ
ϕ = pressure angle, rad

Since the swept volumes we are dealing with all have the same thickness, w, the rate of change of the transverse area is

$$dA = \frac{dV}{w} d\theta = R^2 - \frac{OS^2 + QS^2}{2} \quad \text{in.}^2$$

By applying the law of cosines, this becomes

$$\frac{OS^2 + QS^2}{2} = a^2 + PS^2$$

So the rate of change of the transverse area is

$$dA = R^2 - a^2 - PS^2 \quad \text{in.}^2$$

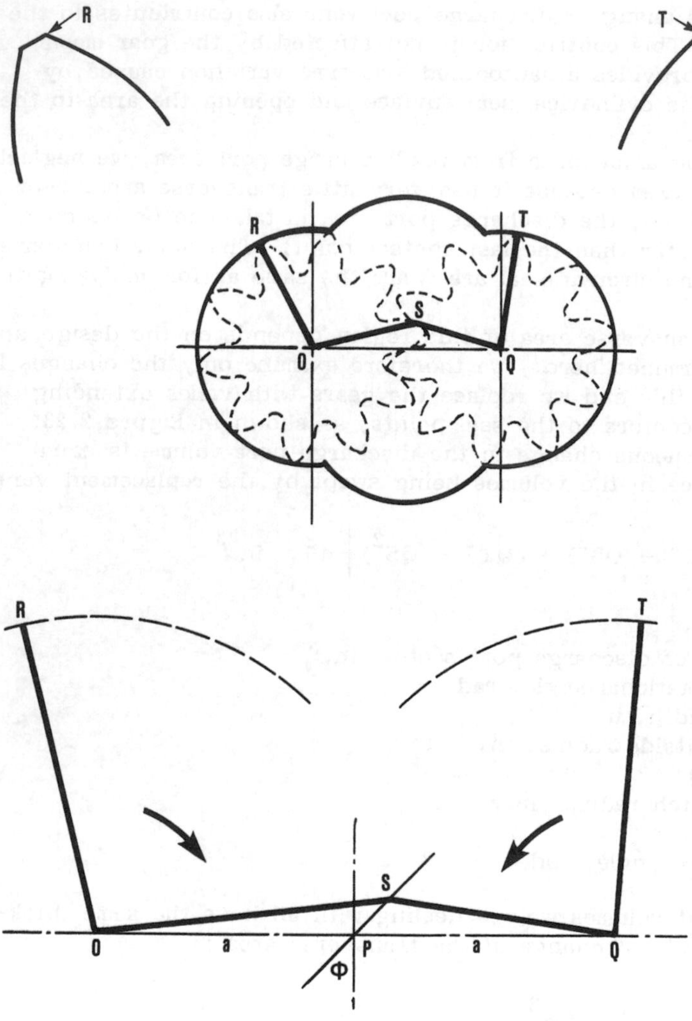

FIGURE 2.23 Discharge port area geometry.

The transverse area is at its minimum when two teeth just contact. When the contact point reaches the pitch point, PS = 0, so the area rate of change is at its maximum. The area is its maximum just as the next pair of teeth start to contact.

The change in axial load, therefore, is proportional to the difference in the areas at the minimum and maximum positions. We find this by integrating the rate of change equation that was just developed. To do this, PS is expressed in terms of

$$PS = a\theta \cos \phi \quad \text{in.}$$

The change in the axial force, then, is

$$\Delta F_A = p\int_{\theta=\alpha}^{\beta} \left[R^2 - a^2 - a^2\theta^2 \cos^2\phi \right] d\theta$$

$$= p\left[(R^2 - a^2)\theta - \frac{a^2\theta^3 \cos^2\phi}{3} \right]_{\alpha}^{\beta}$$

$$= p\left[(R^2 - a^2)(\alpha - \beta) - \frac{(a^2 \cos^2\phi)(\alpha^3 - \beta^3)}{3} \right] \quad \text{psi}$$

The magnitude of the axial force change is found from this equation. The force increases parabolically with time, as the contact point rotates with the gear, and snaps back to the initial value as the next contact occurs. During this cycle another tooth leaves the seal area and adds a pumping chamber to the area under consideration. This adds $2\pi/n$ to the linear $\alpha - \beta$ term. This increment disappears when the next tooth contact is made.

2.3.3 Pressurization Sector Forces

Pressurization of the pumping chambers as they travel from the inlet to the discharge port occurs in two different ways. The first assumes that the pressure increases linearly with rotation over the sector where the chamber is out of communication with the two ports. This is the assumption found in the few books devoted to such analyses, so will be discussed first.

As stated earlier, the length of the cylindrical seal area is determined by the port sizes and gear pitch and is not any particular ratio to the tooth spacing. To aid in the discussion it is assumed the area spans slightly more than five tooth spaces.

When a chamber just seals off from the inlet port, as at A in Figure 2.17, it is still at inlet pressure. Its pressurization is due to leakage from the preceding chamber, over the tip and sides of the preceding tooth. Each preceding chamber is similarly pressurized by leakage from the chamber ahead of it. Assuming that the leakage flow is large in comparison to the very small flow needed to pressurize the chambers, the pressure drops across the teeth are equal. The pressure in any chamber quickly becomes proportional to the number of teeth between it and the discharge port. As shown in Figure 2.17, there are six teeth in the seal area. The pressure in the last chamber to enter the area is at one-sixth the discharge pressure, and the pressures in the chambers that precede it increase in increments of one-sixth of discharge pressure.

As the gear rotates and the next tooth loses contact with the sealing surface, there are four sealed chambers, as shown in Figure 2.17. Since there are only five teeth in the leakage path, the pressure rises in the first chamber from one-sixth to one-fifth of the discharge pressure. When the next chamber is sealed there are again six leakage paths and the chamber pressures quickly rise accordingly, each becoming one-sixth of discharge pressure more than at the start of the cycle.

The magnitude of the radial force due to this pressure profile is the same as the average pressure acting over the circumferential area of the sealed chambers

$$F'_R = \frac{2\pi Rwm}{n}\,\frac{p}{2}$$

$$= \frac{\pi Rwmp}{n} \quad \text{psi}$$

where

R = gear outside radius, in.
w = gear width, in.
m = maximum number of sealed chambers
n = number of gear teeth
p = discharge pressure, psi

The pressure in the most advanced pumping chamber is $mp/(m + 1)$. When this chamber opens to the discharge port area, its load is momentarily lost to the area being considered. This lost load is equal to

$$\Delta F'_r = \frac{2\pi Rwmp}{n(m + 1)} \quad \text{lb}$$

so the load for this moment is

$$F''_r = \frac{2\pi Rwp[(m/2) - m/(m + 1)]}{n}$$

$$= \frac{2\pi Rwpm(m - 1)}{2n(m + 1)} \quad \text{lb}$$

This drop in load occurs even if the pump geometry is such that a new chamber seals as one opens to the discharge port. In this case the load quickly returns to the previously determined level. If the two events are staggered, the load rises to the equilibrium level for one less chamber. This is

$$F'''_r = \frac{\pi R w p(m - 1)}{n} \quad \text{lb}$$

The effect of the pressure staircase is nearly the same as if the pressure increased linearly along the periphery of the gear, being zero at the last tooth to enter the sealing area and full discharge pressure at the last tooth in the sealing area. The centroid of this triangular pressure distribution, then, is at the third point from the high end. This is at a point 5/3 tooth spaces from the last tooth in the sealing area when there are five chambers. It rotates with the gear until a tooth leaves the seal area. There are then four chambers and the centroid shifts to the point 4/3 from the high-pressure end. As a new tooth enters the sealing space from the inlet port, the centroid reverts back to its starting point. This cycling occurs at pumping frequency

$$f_p = \frac{n(\text{rpm})}{60} \quad \text{Hz}$$

where

n = number of gear teeth
rpm = shaft speed, rev/min

Because the gears mesh, the tooth on one has the same timing as a space on the other. Load and centroid changes on one gear are therefore out of phase with those of the other by one-half a tooth spacing. In terms of pumping frequency they are 180° out of phase.

Because of leakage, the pressures in the oil film on the transverse surfaces of the teeth are compatible with those of their adjoining pumping cavities. The pressure profile on the transverse surfaces duplicates that on the circumferential surfaces. The axial force in this sector, then, arises from pressure in an annular area extending from the root to the outside radius of the teeth. This axial force is equal to

$$F'_A = \frac{mp}{n} \frac{\pi}{4} (R^2 - r^2)$$

$$= (R^2 - r^2) \frac{mp\pi}{4n} \quad \text{lb}$$

Since the pumping chambers have constant transverse areas, the axial force centroid has identically the same characteristics as the radial force centroid, which also results from constant areas.

The angular distance from where a pumping chamber becomes sealed from the inlet to where this chamber returns to the inlet port is much greater than 180°. Reaching the discharge pressure early in this distance provides pressure to balance some of that in later sectors so that the gear shaft bearing loads will be reduced. This is shown later in the section on bearing loads.

The bearings in modern high-pressure pumps are critical, so this load reduction is important. Side plates like that shown in Figure 2.22 are used to facilitate the load reduction. The chambers leading from the discharge port area, around the periphery of the seal area, short circuit the leakage paths around the teeth. Actual measurements show that pumping chamber pressure reaches discharge pressure within a few degrees after the chamber seals off from the inlet port.

This device increases the leakage since the leakage path now is around one or two teeth instead of four or five. However, these same pumps are equipped with pressure-loaded side plates that minimize axial clearance. Further, the gears are made with a light press fit in the housing so that they scrape or burnish the housing to their exact outside diameter. Apparently, these measures keep the leakage within acceptable limits.

For this analysis, as in those of piston and vane pumps, we assume that the pressure in a pumping chamber instantly rises to the discharge pressure as soon as the chamber seals off from the inlet port. The radial force of all the chambers in the sealing area, then, is

$$FF_r = \frac{2\pi Rwmp}{n} \quad \text{lb}$$

In our example pump where the sealing area spans slightly more than five tooth spaces, there are either four or five sealed chambers at a time. The radial force range is found by calculating the loads with the equation above, using m = 4 and 5.

The centroid of the radial load is located at the center of the sealed areas. It starts 2.5 tooth spaces from an incoming tooth as it enters the seal area and rotates with the gear until a chamber opened to the discharge port. The centroid at this time jumps to two spaces from either end and continues to rotate with the gear until the next tooth entered the seal area.

This force is twice what it would be if the pressure built up slowly, and the housing must be designed to carry it. As we see later, it will, when combined with the load in the high-pressure port, result in a net reduction in bearing load.

2.3.4 Torque

We return to the discharge port to analyze the torque. There, pressure acts against the last tooth in the cylindrical pressurization sector to resist the gear motion. Pressure acting against the last teeth in contact assists the motion. The net torque, then, is due to the differences in the active lengths of the teeth in the two areas. The analysis is simplified if the gears are replaced with vanes like those shown in Figure 2.23. In this analysis we are concerned with the last teeth in contact, while in the previous analysis, the first teeth in contact were considered. However, since we have identical gears, the geometries in the two zones are symetrical to each other. The sequence of the cyclic changes are reversed, but the cyclic ranges are the same. Using the geometry of Figure 2.23, the torque is

$$T = \frac{pw}{2}[(OR^2 - OS^2) + (QT^2 - QS^2)] \quad \text{in.-lb}$$

Following the same mathematical development, the torque is

$$T = pw[R^2 - b^2 - (a^2\theta^2 \cos^2\phi)] \quad \text{in.-lb}$$

Torque is highest when the contact is at the pitch point and declines as the contact moves from this point. Torque is lowest at the end of the tooth contact as defined by the previously found angle, α. The torque range is found by taking the difference in the torque at this angle and that when the tooth contact is at the pitch point,

$$\Delta T = pwa^2\alpha^2 \cos^2\phi \quad \text{in.-lb}$$

2.3.5 Bearing Loads

From a noise generation standpoint, analysis of the bearing loads is redundant since the major part of these forces comes from loads that have already been analyzed. The purpose of including them is to broaden our understanding of pump force systems. We can meet this objective with rather simple analyses.

The difference between the previously calculated radial loads and the bearing load is easily illustrated by considering pressure acting in a tube whose cross section is a semicircle closed by a straight side. According to the previous analyses, the loads on the two sections are equal to the product of the pressure times their area. Because of this there is π times more load on the curved surface than on the flat surface. Yet their loads are equal since the stresses at their junctions are at equilibrium.

There are several ways to evaluate the load on the semicircular surface. If the concern is the stress or deflection of the curved structure, the method mentioned above is appropriate. It is the equivalent of summing or integrating the pressure loads on incremental areas of the total area.

When the concern is the effect of the load external to the structure, the directions of the incremental loads must be considered. This is done by separately summing orthogonal components of the incremental loads. In the case of the semicircular surface, the appropriate orthogonal directions would be parallel and perpendicular to the flat face. The summation of the parallel components of one-half the cylindrical surface cancels that from the other half, so the total in this direction is zero. The summation perpendicular to the flat face equals the pressure force on the flat face.

No matter what the shape of the pressurized surface, the resultant of summing the orthogonal components of its incremental loads equal the product of the pressure times the areas created by projecting the surface onto planes that are perpendicular to the orthogonal directions. Where one of the projected areas spans the entire pressurized surface, the sum of the loads on the areas perpendicular to it will always be zero. The entire external load of the pressurized surface, then, is simply pressure times the spanning projected area.

Returning to the bearing loads, we first consider the case where there is a pressure gradient along the path from inlet to discharge. Ernst (6) analyzed this case using the summation process. He assumed that the pressure gradient extended around the gear for 180° and that the discharge pressure acted over an additional 90°, as shown in Figure 2.24. He found that the bearing force was equal to

$$F_b = 1.635Rwp \quad \text{lb}$$

where

R = gear outside radius, in.
w = gear width, in.
p = discharge pressure, psi

To examine the effect of achieving full pressure as soon as the chambers seal to the inlet port, we assume that the same pressurized sectors are at full discharge pressure, as shown in Figure 2.24. The easiest way to evaluate the resultant of this pressure distribution is to consider the pressure as acting only on the projected area. In Figure 2.24 this area is represented as the line spanning the unpressurized quadrant. On this basis the resultant is

$$F_b = 2Rwp \sin \frac{\pi}{4}$$
$$= 1.41Rwp \quad \text{lb}$$

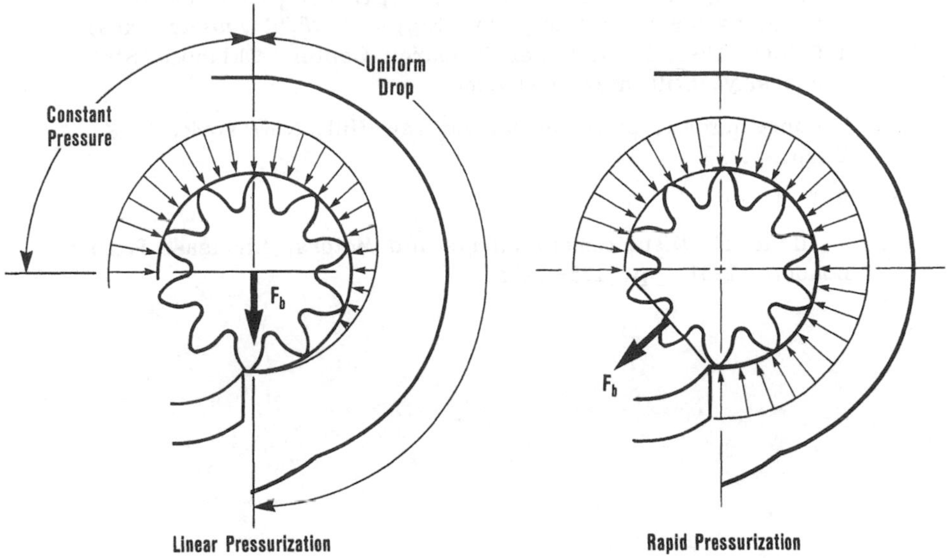

FIGURE 2.24 Gear pump bearing loads.

This analysis shows that early pressurization reduces the bearing loads by about 14%. A more detailed analysis would show a larger reduction.

An excellent detailed analysis of gear pump bearing loads is made by Hadekel (7). He is very ingenious in his application of projected areas in this analysis, and a review of this work is recommended to persons with a keen interest in analysis techniques.

Reactions to the gear tooth loads add to these pressure-induced bearing loads. For reasons already given, an analysis of these loads is not given here. Hadekel is also recommended to those interested in such detail.

REFERENCES

1. W. Ernst, *Oil Hydraulic Power and Its Industrial Applications*, 2nd ed., McGraw-Hill, New York, 1960, pp. 142–144.
2. G. A. Fazekas, "On Half Harmonics," ASME paper 70-WA/DE 16, American Society of Mechanical Engineers, New York, Nov. 1970.
3. J. U. Thoma, *Modern Oilhydraulic Engineering*, Trade and Technical Press, Morden, Surrey, England, 1970, p. 134.

4. R. K. King and D. C. Headrick, "Optimizing Gear Pump Parameters for Decreasing Flow Ripple," *BFPR Journal* 14(2): 149–153, 1981, Fluid Power Research Center, Oklahoma State University, Stillwater, Oklahoma.

5. E. Buckingham, *Spur Gears*, McGraw-Hill, New York, 1928, Chap. 2.

6. Ref. 1, pp. 79–80.

7. R. Hadekel, *Displacement Pumps and Motors*, Sir Isaac Pitman, London, 1951, pp. 125–129.

3
Vibration Basics

Both airborne and structureborne noise are the result of structural vibrations. It is interesting to note that many textbooks on acoustics devote most of their pages to the subject of vibration. Such depth is not necessary for reducing machine noise. Only a general familiarity with the nature of vibrations is provided here. Some of the information from this chapter is applied and expanded in Chapters 4, 9, and 10.

3.1 STRUCTURAL RESPONSE

Structural deflections are constant or varying, depending on the nature of the force causing them. Constant deflections do not generate noise. Noise is generated by motion, often by vibrations in the microinch range.

Structural responses to forces are frequency dependent. For this reason we find it convenient to analyze vibrations as though they occur at only one frequency. Later we deal with the fact that vibrations can consist of many frequencies.

3.1.1 Sinusoidal Motion

When a structure vibrates at a single frequency it is said to have simple harmonic motion. This motion is also described as sinusoidal because a time plot of its amplitude, velocity, or acceleration is a sine curve. For the same reason we sometimes refer to these motion parameters as waves.

Frequency, of course, is the number of times a vibration cycle repeats each second. The unit for cycles per second is the hertz.

Since sine waves are a function of angles, we also have an angular or *circular frequency* ω

$$\omega = 2\pi f \quad \text{rad/sec}$$

$$= 360f \quad \text{deg/sec}$$

where f is the frequency in hertz. The radian form is the most generally used because it facilitates mathematical analysis. Equations for vibrations are usually in the form

$$\sin \omega t \quad \text{or} \quad \cos \omega t$$

where t is the time in seconds.

The *period* of a vibration is the time for one complete cycle

$$T = \frac{1}{f} \quad \text{sec}$$

Since the sine function completes its cycle in 360° or 2π rad, the period is also sometimes given these angular values.

There is a simple relationship between the deflection, velocity, and acceleration of a single-frequency vibration. If the instantaneous displacement is expressed as

$$x = a \sin \omega t \quad \text{in.}$$

where a is the maximum deflection, measured from the undeflected position, in inches. The instantaneous velocity is found by differentiating this equation

$$\dot{x} = \frac{dx}{dt}$$

$$= \frac{d(a \sin \omega t)}{dt}$$

$$= \omega a \cos \omega t \quad \text{in./sec}$$

Differentiation of this velocity equation provides the instantaneous acceleration

$$\ddot{x} = \frac{d\dot{x}}{dt}$$

$$= \frac{d(\omega a \cos \omega t)}{dt}$$

$$= -\omega^2 a \sin \omega t \quad \text{in./sec}$$

It should be noted that each differentiation increases the amplitude of the resulting wave by a factor equal to the circular frequency. Also, the acceleration is a sine function like the displacement but has a negative sign. It therefore decreases at times when displacement increases.

3.1.2 Free Vibration

Pressurized surfaces of all hydraulic components, pump structures carrying pumping forces, and plates in machine structures are all examples of vibration systems. The basic requirements for vibration are fairly simple. There must be a mass that is restrained from moving by an elastic element whose resistance increases with displacement. There must also be an external force. Systems also have damping that resists motion by converting mechanical energy to heat. Typically, this is provided by friction in joints, material hysteresis, or fluid viscosity. Figure 3.1 shows a simple system meeting these specifications. The vibration characteristics of this system are typical of those of many structures.

When the mass in Figure 3.1 is displaced, then suddenly released, the mass begins to oscillate. This is called *free vibration*. If the damping, provided in this system by a dashpot, is relatively low, the mass will vibrate at the system *natural frequency*

$$f_n = 0.159 \sqrt{\frac{K}{M}} \quad \text{Hz}$$

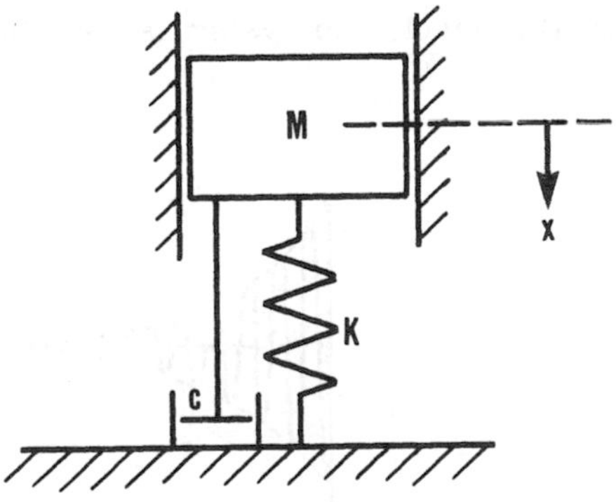

FIGURE 3.1 Simple vibration system.

where

K = spring rate, lb/in.
M = mass, lb-sec^2/in.

Displacing the mass added potential energy to the system equal to

$$PE = \frac{Kx_0^2}{2} \quad \text{in.-lb}$$

where x_0 is the initial displacement in inches. When the mass is released and begins to move, this energy begins to convert into kinetic energy. The conversion is complete when the undisturbed position is reached. The kinetic energy at this point is

$$KE = \frac{M\dot{x}^2}{2} \quad \text{in.-lb}$$

where $\dot{x}$ is the velocity of mass at x = 0.

This energy converts back and forth as the mass moves through its cycle. The system energy at any instant is the sum of the two forms of mechanical energy. In a lossless system the total energy remains constant. However, in real systems damping slowly converts the mechanical energy in the system to heat. Consequently, the amplitude of the oscillations slowly decreases, as shown in Figure 3.2. This behavior—a sudden onset, then decaying oscillations at the system natural frequency—is typical of structures subjected to impact or other shocks.

The oscillations decay more rapidly as the restraint of the dashpot is increased. Their frequency also decreases slightly when this is done. The damping can be increased to a point where the disturbed mass just creeps back to its equilibrium position, without vibration. When the dashpot resistance reaches this level, the system is said to

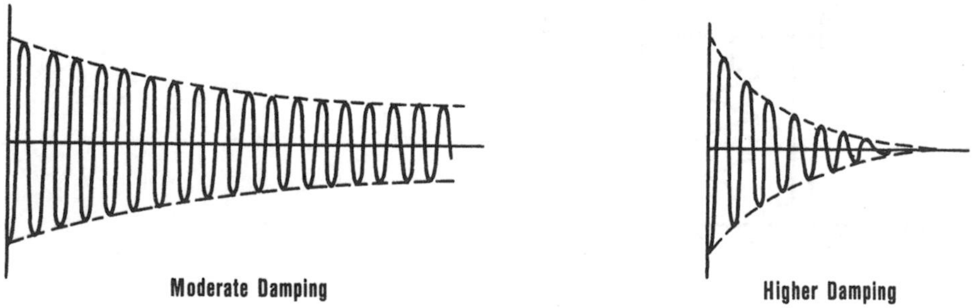

FIGURE 3.2 Damped vibration.

be critically damped. Critically damped systems seldom occur and are of interest only because they provide a benchmark for judging lower damping levels.

The frequency of a freely vibrating system that has significant damping is called the *damped natural frequency* and is equal to

$$f_d = f_n\sqrt{1 - C} \quad \text{Hz}$$

where

C = ratio of damping to critical damping
$= c/c_c$
c = damping, lb-sec/in.
c_c = critical damping, lb-sec/in.

3.1.3 Forced Vibration

If a sinusoidially varying force is applied to the mass instead of a single perturbation, a sustained vibration results. This vibration is at the frequency of the applied force and is called a *forced vibration*. Its amplitude is a function of the ratio of the force frequency and the system natural frequency. If the applied force is

$$P = p \sin \omega t \quad \text{lb}$$

where

P = force amplitude, lb
ω = force circular frequency, rad/sec

the instantaneous displacement is

$$x = \frac{P/K}{(1 - F^2)^2 + (2CF)^2}$$

where

F = ratio of forcing to natural frequency
$= f/f_n$
$= \omega/\omega_n$

It should be noted that the numerator of this equation is the deflection that occurs if the sinusoidially varying force acts against just the spring. The denominator provides the effect of the spring-mass system and the damping. It can be thought of as a dynamic factor that amplifies or attenuates deflections caused by the force acting on the spring.

$$DF = \frac{1}{(1 - F^2)^2 + (2CF)^2}$$

In Figure 3.3 this dynamic factor is plotted against the ratio of the forcing and natural frequencies, F, for several damping ratios, C. The most notable feature of this plot is that the amplitude of forced vibrations are amplified when the forcing frequency and the natural frequencies are nearly equal ($F \approx 1$). Maximum amplification occurs at the *resonant frequency*. This frequency is slightly less then the natural frequency and the difference increases as damping increases.

It is also obvious from Figure 3.3 that increasing damping decreases the amplification at resonance. To provide a frame of reference, systems with amplifications below about three are of little practical interest. Systems that have amplifications above 100 are rare. From this it is concluded that we are interested in systems having damping ratios in the range $0.005 < C < 0.17$.

Three similar terms have been discussed: natural frequency, damped natural frequency, and resonant frequency. Failure to recognize the differences in them can lead to confusion in reading vibration literature. However, there are no practical differences in the numerical values of these quantities, in the range of damping ratios that was just given. For this reason natural frequency and resonant frequency are often used interchangeably.

When the forcing frequency is low in comparison to the natural frequency, the dynamic factor is nearly 1. This indicates that most

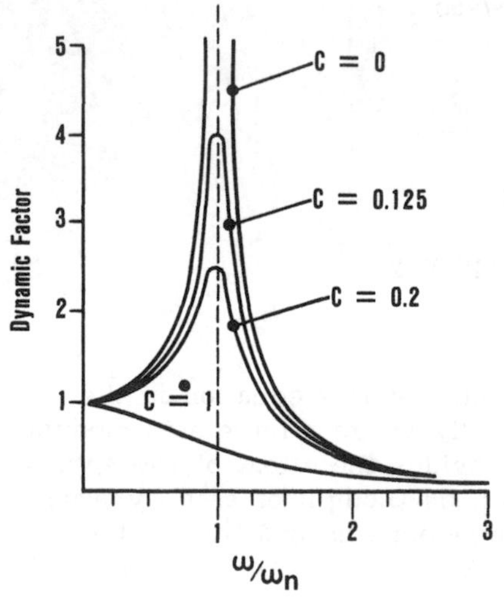

FIGURE 3.3 Forced vibrations dynamic factor.

of the resistance to the force comes from spring deflection. The effect of the mass is proportional to acceleration and damping tends to be proportional to the velocity, both being relatively low at low frequencies. For this reason the range of frequencies below about six-tenths of the natural frequency is sometimes referred to the *spring-controlled region.*

As we saw earlier, velocity increases linearly with frequency, relative to deflection, and acceleration increases with the square. At resonance, the spring and mass reactions are equal and, as indicated by their signs in the equation, they oppose each other. The applied force is then principally resisted by the damping, and this accounts for the strong influence that damping has at resonance.

At frequencies well above resonance the acceleration becomes quite large relative to deflection and velocity. The mass then offers the principal resistance to the applied force. Because of this the frequencies above about $1.4f_n$ are sometimes referred to as the *mass-controlled region.*

Increasing damping is often thought to be a key vibration control. As we have seen, it has a strong influence only when there is a resonant condition. It does reduce vibrations in the mass-controlled region, but these are already attenuated so a further decrease is of limited value. Damping also increases the decay of free vibrations, so it is useful in the rare cases where vibrations are the result of periodically occurring impacts. In general, however, damping is seldom effective in noise reduction.

3.1.4 Degrees of Freedom

The simple vibration system that we have been discussing is a single-degree-of-freedom system. This means that it can vibrate in only one mode and it provides a very useful model for understanding many of the vibrations that we encounter in practice. Unfortunately, not all vibrations occur in single-degree-of-freedom systems.

Multiple masses, connected resiliently, vibrate in many modes and are called multidegree-of-freedom systems. Figure 3.4 shows a system of this type. Since each of the masses can move independently of every other mass, it is a two-degree-of-freedom system.

The cantilevered beam shown in Figure 3.4 is also a multidegree-of-freedom system. Three of its vibration modes are shown, but others are possible. Because of this ambiguity it is difficult to characterize the degrees of such systems.

Large plates or sheet metal panels are a common example of multimode or multidegree-of-freedom systems. Figure 3.5 shows some of the modes of these members.

At low frequencies the whole plate acts like a diaphragm, with all parts vibrating in unison. At higher frequencies the plate divides into a number of active areas or panels bounded by nodal lines where

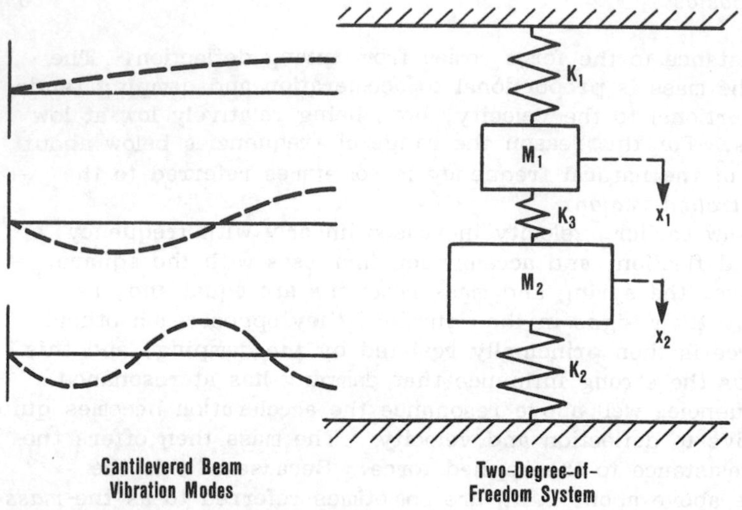

FIGURE 3.4 Multidegree-of-freedom systems.

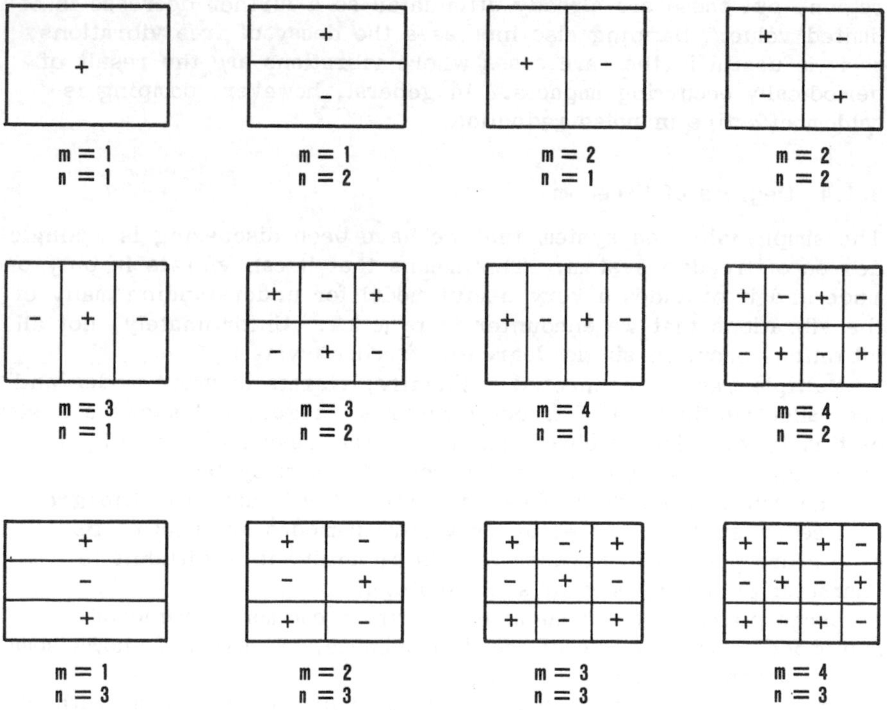

FIGURE 3.5 Some plate vibration modes.

no vibration takes place. These are the rectangular areas on the plates shown in Figure 3.5. Each area acts like an individual simply supported plate. As one panel deflects upward the adjacent panels deflect downward, as indicated by the symbols. The number of these elemental panels increases with frequency.

Plates are capable of vibrating in so many modes that a two-number system is used to categorize them; each number referring to the number of vibrating panels along each dimension. With this system the individual panels can be thought of as vibrating in their 1, 1 mode.

The resonant frequency associated with each vibration mode is calculated from

$$f_{(m,n)} = 368t\sqrt{\frac{E}{w}}\left(\frac{m^2}{a^2} + \frac{n^2}{b^2}\right) \quad \text{Hz}$$

where

E = modulus of elasticity of plate material, psi
t = plate thickness, in.
w = weight of one cubic foot of plate material, lb/ft^3
m = number of vibrating panels along the length
n = number of vibrating panels along the width
a = length, in.
b = width, in.

3.2 WAVE ANALYSIS

The use of the wave analogy to describe forces and vibrations provides us with a number of convenient tools. Most of these are based on the characteristics of the sine function.

3.2.1 Sine Basics

A sine wave, as shown in Figure 3.6(a), is symmetrical about zero; therefore, its time average is zero, regardless of its amplitude. Because of this it cannot be measured with conventional instrumentation, so other ways to describe it have developed. Their use depends on how the resulting statistic will be used.

The simplest measure of a sine curve is made by flipping the negative halves upward as shown in Figure 3.6(b). Electrically, this is done with a full-wave rectifier. The average of this modified wave is 0.636 times the peak value of the sine wave. This measure of the wave finds its greatest use in vibration measurements.

Sound measurements are generally based on the mean square of the sine wave. This is the average when the wave is squared as shown in Figure 3.6(c).

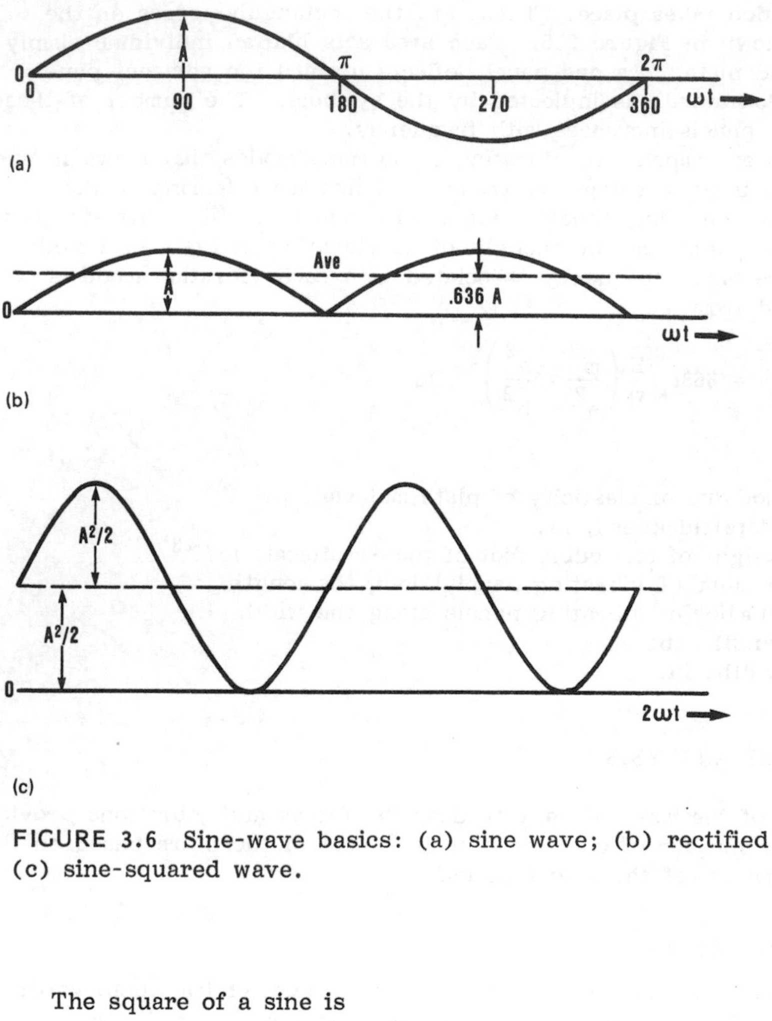

FIGURE 3.6 Sine-wave basics: (a) sine wave; (b) rectified sine wave; (c) sine-squared wave.

The square of a sine is

$$(A \sin \omega t)^2 = \frac{A^2}{2} - \frac{A^2}{2} \cos 2\omega t$$

This function cycles from zero to A^2 and its average is half its height, $A^2/2$. This statistic is especially important because it is a measure of the energy in the original sine wave.

More frequently, the square root of this quantity is used. This is the ubiquitous root mean square (rms) found in sound and vibration literature

$$\text{root mean square} = \frac{A}{\sqrt{2}}$$

$$= 0.707A$$

where A is the sine-wave amplitude, or peak value.

3.2.2 Adding Sine Waves

When two sine waves of the same frequency are combined, their sum is a sine wave of the same frequency. If they both pass through their zeros and maxima at the same time, they are said to be in phase and their combined amplitude is the arithmetic sum of their amplitudes.

When the two waves do not pass through identical events at the same time, they have a phase difference. The time difference in their passing through the cyclic events, say their maxima, is measured in terms of an angle, the time for one cycle being equal to 360° or 2π rad. This angular difference is termed the *phase angle*.

The simplest way to analyze how such waves combine is to use vectors (1,2). The amplitude of the combined wave is the sum of two vectors having lengths proportional to the two wave amplitudes and with the angle between them equal to the phase angle, as shown in Figure 3.7. Further, the angle that the resultant makes with the individual wave vectors provides the phase angles between the combined wave and the originals.

From Figure 3.7 it is seen that the amplitude of the combined wave is largely a function of the phase angle between the original two waves. When this angle is zero, the amplitude of the resultant is the algebraic sum of the amplitudes of the original waves. When the waves are 180° out of phase, the resultant is the algebraic difference of their amplitudes. The resultant amplitude is between these two extremes for intermediate phase angles.

Perhaps this is easiest to visualize when the two waves have equal amplitudes. When the two are in phase, their resultant has twice their amplitude. When they are 180° out of phase, they cancel each other.

Later, in analyzing pump noise we will be interested in the addition of any number of equally spaced equal-amplitude sine waves of the same frequency. As in the case of adding two waves, the amplitude of the resultant of adding many equal-frequency waves is the sum of vectors representing the amplitudes and phase differences of the waves. Figure 3.8 shows the vectors for five equal-amplitude, equally spaced waves. From this figure it is apparent that whenever equal waves are also equally spaced, they cancel.

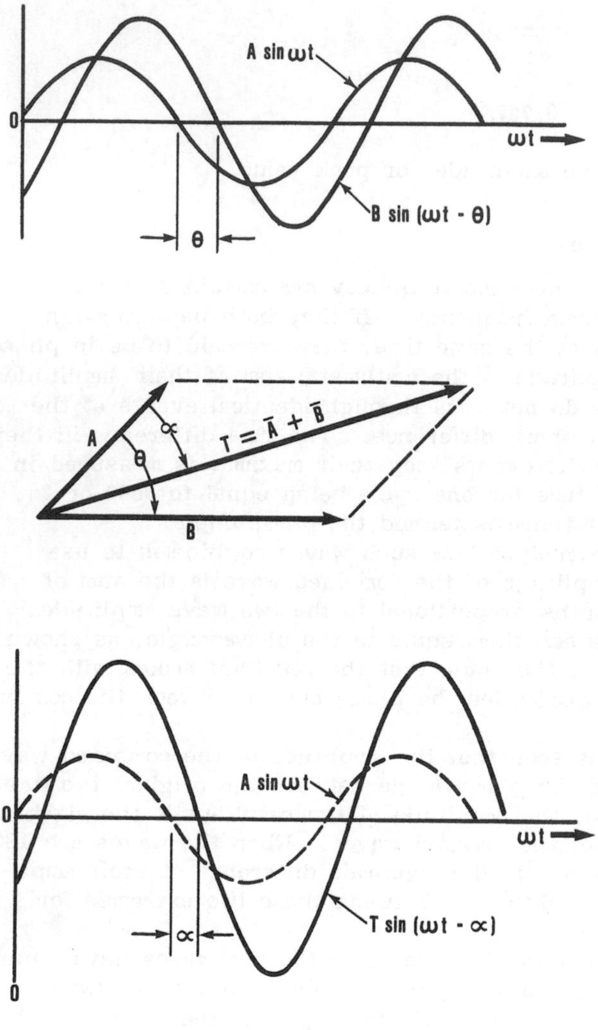

FIGURE 3.7 Vector addition of sine waves.

Although this vector method of determining the sum of sine waves strictly holds only for waves of the same frequency, it is bent a bit and used to analyze the effect of adding two waves with slightly different frequencies (3). Because of the frequency differences, the phase angle between the waves will change constantly. At some time

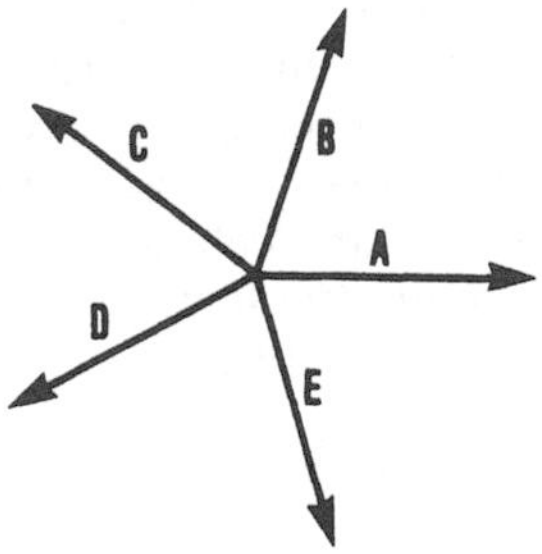

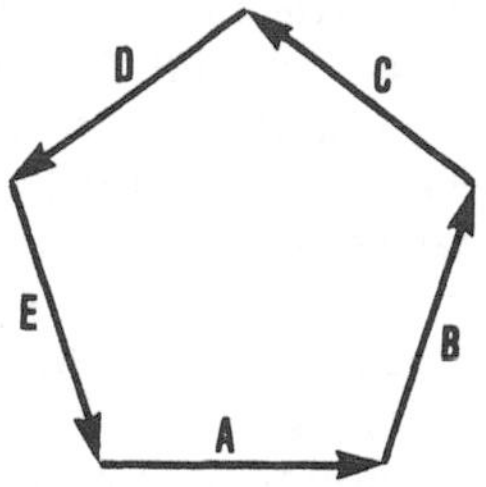

Closed Figure Indicates
That the Resultant Is Zero

FIGURE 3.8 Vector sum of five equal and equally spaced sine waves.

they are in phase for a few cycles and their amplitudes add. At some other time they are 180° out of phase and their combined amplitude is the algebraic difference in their individual amplitudes. Their total amplitude, then, continuously varies between these two limits at a frequency equal to the difference in their two frequencies. When this occurs it is called a *beat note*.

3.2.3 Complex Waves

Forces and vibrations found in hydraulically powered machines are rarely of the single-frequency type. Although they may have complex variations with time, we refer to them as waves as long as they are periodic. All such periodic waves are actually a combination of a series of sine waves with frequencies equal to or multiples of the complex wave frequency. This series of sine waves is called a *Fourier series*. Any periodic wave that is a function of time can be expressed in a Fourier series as (4)

$$f(t) = A_0 + A_1 \sin(\omega t + \theta_1) + A_2 \sin(2\omega t + \theta_2) + A_3 \sin(3\omega t + \theta_3) + \cdots$$

where

f = is the frequency of the complex wave, Hz

$\omega = 2\pi/f$, rad/sec

A_1, A_2, A_n = amplitudes of waves in the series

$\theta_1, \theta_2, \theta_n$ = phase angles of these waves

The first term is the time average of the complex wave. This will sometimes be omitted in this book because it does not contribute to noise. The second term has the same frequency as the complex wave and is called the *fundamental frequency*. The subsequent terms have frequencies that are multiples of the fundamental and are called *harmonics*. They are numbered by how many times their frequency exceeds the fundamental. The wave represented by the term $A_3 \sin(3\omega t + \theta_3)$, for example, is the third harmonic. Computing or experimentally determining the Fourier series of a complex wave is called *harmonic analysis*.

In this chapter equations for vibrations are in terms of frequency. In Chapter 2 we saw that loads which cause these vibrations are nominally determined as functions of time. The Fourier series provides the very important bridge between frequency and time functions. Some books on Fourier analysis express this as the ability to transfer between the *frequency domain* and the *time domain*. Such transfers are illustrated in later sections of this chapter.

The Fourier series is also sometimes written

$$f(t) = a_1 \sin \omega t + a_2 \sin 2\omega t + \cdots + a_n \sin n\omega t + \cdots + b_1 \cos \omega t + b_2 \cos 2\omega t + \cdots + b_n \cos n\omega t + \cdots$$

The coefficients a and b are determined from the following integrations:

$$a_n = \frac{\omega}{\pi}\int_0^{2\pi/\omega} f(t) \sin n\omega t \, dt$$

and

$$b_n = \frac{\omega}{\pi}\int_0^{2\pi/\omega} f(t) \cos n\omega t \, dt$$

Where equations for the wave can be written, algebraic integration is used to determine these coefficients. This tends to be a tedious but not a very difficult process. Fourier analysis is so widely used that numerous shortcut techniques have developed and are found in mathematics and electrical engineering textbooks (5).

When the wave function is known but cannot be expressed in integrable terms, numerical integration is used. This requires determination of the wave amplitude for small, equal increments of time. When this is done, the number of increments must equal twice the number of significant harmonics present in the wave (6). For noise studies, frequencies above 2000 Hz are seldom important. On this basis it is safe to assume that the number of significant harmonics is equal to 2000 divided by the fundamental frequency of the wave being analyzed. Computer programs for determining the harmonic coefficients from such data are commonly available.

In noise studies we want to know the total strength of each harmonic, so we combine the sine and cosine terms for each frequency. The cosine function lags the sine by 90°, so the combined amplitude is found by adding two vectors at right angles to each other. Algebraically, this is

$$A_n^2 = a_n^2 + b_n^2$$

The phase angles are seldom useful. However, if wanted, they can be determined from

$$\tan \theta_n = \frac{a_n}{b_n}$$

3.3 EFFECT OF WAVE SHAPE

Controlling wave shapes is one of the most important ways of reducing noise. The general effects of such changes are best illustrated by utilizing formulas in handbooks for determining the harmonic components of simple wave shapes (7). In Chapter 5 we discuss the harmonic content of more realistic wave shapes.

Figure 3.9 shows a square wave and its harmonic analysis. In Chapter 2 we saw that many of the pump force fucntions are idealized by this wave.

The logarithms of the harmonic amplitudes or strengths are plotted because this scaling is compatible with the method used in rating noise. Typically, the amplitudes become smaller as the harmonic number increases. The fact that only the odd harmonics are present is uncommon but does occur with other simple wave shapes.

The effect of changing wave shape on the harmonic content is best illustrated by comparing the Fourier series of a square wave with that of a half-sine wave, shown in Figure 3.10. The comparison is also shown in this figure. From this comparison we generalize that "rounding" the corners of a wave greatly reduces the strength of its harmonics.

Perhaps this comparison suffers from the fact that a square wave has only odd harmonics and a half-sine wave has only even harmonics. Because of this, no conclusions can be drawn for the effect on the fundamental of smoothing the wave. In general, there will be very little effect on this component.

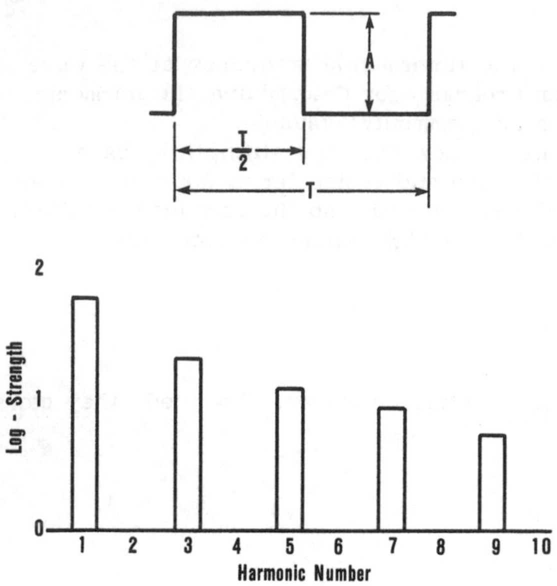

FIGURE 3.9 Square wave and its harmonic components.

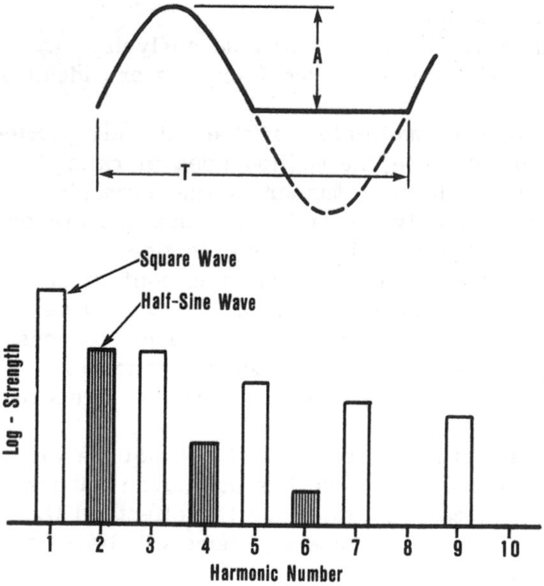

FIGURE 3.10 Half-sine wave and its harmonics.

This comparison was made to demonstrate clearly the effect of wave shape changes by using an extreme example. A change this great is not possible in practice. We now look at more reasonable modifications.

Square waves are generated by instantaneous changes in pressure. When steps are taken to slow these changes the square wave becomes trapezoidal as shown in Figure 3.11. Two transition times x are analyzed. Because of their similarity to a square wave, these trapazoids also have only odd harmonics. Only small reductions in the harmonics occurred when the transition time was increased to one-twentieth of the period. Doubling this time produced much larger reductions.

Another possible modification to a square wave is to change its duration. The effect of this change is shown in Figure 3.12. Two different durations Y are analyzed. Reducing the duration produced components at all harmonics except those divisible by 5. Although the rectangular waves have more components, these decreased more rapidly than those of a square wave, which are also shown in this figure.

The duration effect is not consistent. Some harmonics of the shortest duration wave, Y = 0.3T, are higher than the same harmonics of a wave having a duration of Y = 0.4T.

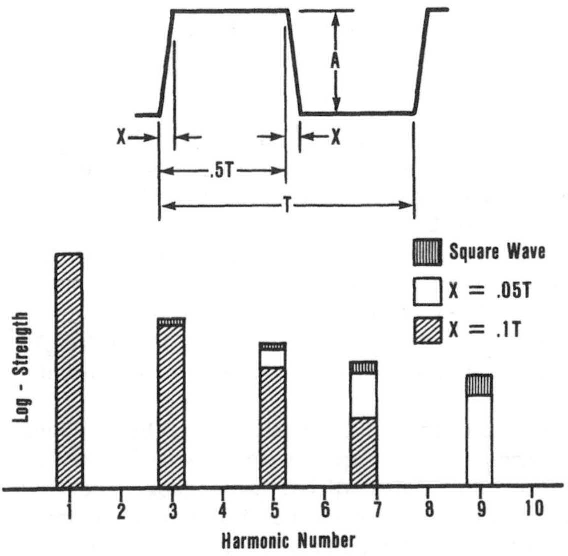

FIGURE 3.11 Trapezoidal waves and their harmonics.

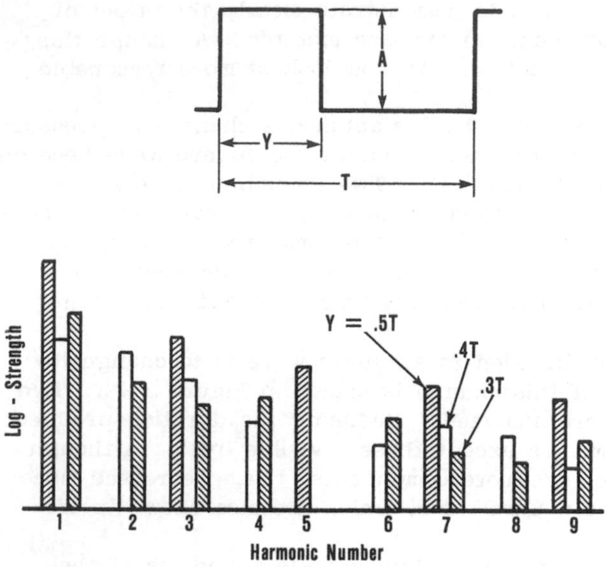

FIGURE 3.12 Effect of varying pulse width.

The effect of increasing duration was also analyzed. It was found that the effect was identical to decreasing duration. The change increment apparently was the controlling factor, as Y = 0.4T had identically the same harmonics as Y = 0.6T.

Additional analyses are made of sawtooth waves. These waves are shown in Figure 3.13. The true sawtooth wave is modified by increasing the transition time from zero to 0.1T. The amplitude of all waves analyzed so far have the same height. A lower height was assigned to the modified sawtooth because it seemed consistent with the other change.

Sawtooth waves have components at all harmonics. These decreased at about the same rate as those of the square wave, which are also shown in Figure 2.13. Those for the modified wave decreased so rapidly, however, that they do not show above the fifth harmonic. It should be noted that although this modification is very effective in reducing the strength of the higher harmonics, it does not change the strength of the fundamental.

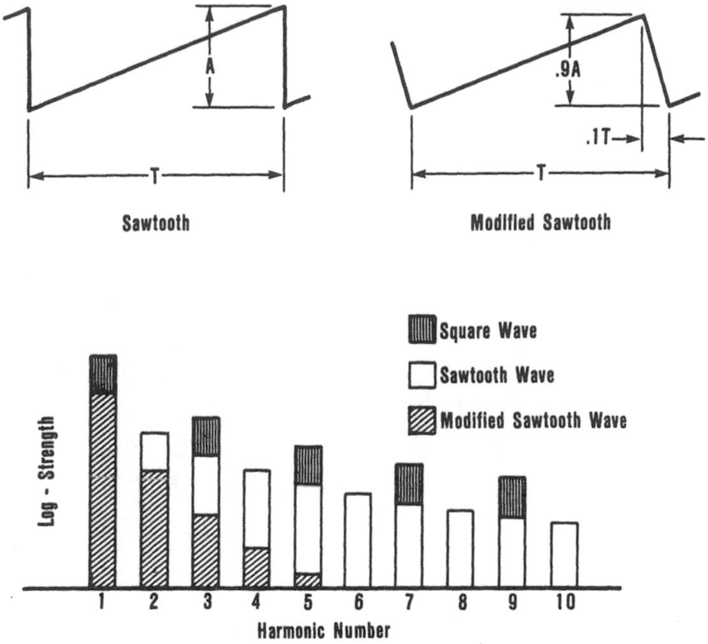

FIGURE 3.13. Effect of sawtooth wave modification.

REFERENCES

1. J. P. Den Hartog, *Mechanical Vibrations,* 4th ed., McGraw-Hill, New York, 1956, Sec. 1.2.
2. R. J. Manley, *Waveform Analysis,* Chapman & Hall, London, 1950, pp. 14–19.
3. Ref. 2, pp. 19–28.
4. Ref. 2, pp. 144–151.
5. Ref. 2, pp. 158–163.
6. S. A. Hovanessian, *Computational Mathematics in Engineering,* D. C. Heath, Lexington, Mass., 1976, pp. 172–173.
7. H. T. Kohlhass, *Reference Data for Radio Engineers,* Federal Telephone and Radio Corp., New York, 1943.

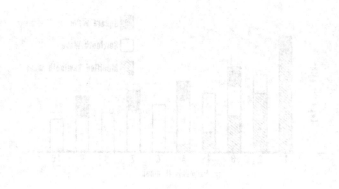

4
Sound Basics

We have already reviewed the mechanics of vibrations that produce noise and the forces that drive these vibrations. Now we examine the mechanics of sound: how it is generated, radiated, reflected, rated, and heard.

4.1 NATURE OF SOUND

As in mechanical vibrations, we simplify sound studies by considering single-frequency sounds. In part this is possible because the Fourier theorem lets us separate complex excitations into a series of single frequencies. In his book on electrical networks, Stanley (1) gives the other part, elegantly: "The response of the complex input is the sum of the responses that would arise from the individual inputs." The behavior of sound is sensitive to the ambient pressure. Sound pressures are so miniscule in comparison to this pressure that even the very highest sound levels encountered in practical experience do not change this behavior. For this reason the principle of superposition is valid for sound as it is in many fields of mechanics.

Perhaps the best way to visualize the nature of sound is to consider how it is generated by an idealized noise source, such as the one shown in Figure 4.1. The piston is driven back and forth by a crank whose speed is in the acoustic range, from 100 to 8000 revolutions per second. Because of the scotch yoke mechanism the piston has sinusoidal motion. The tube is assumed to be infinitely long, so sound waves dissipate before they reach the end. This assumption is made because at this time we do not wish to deal with the complications of reflections from the end.

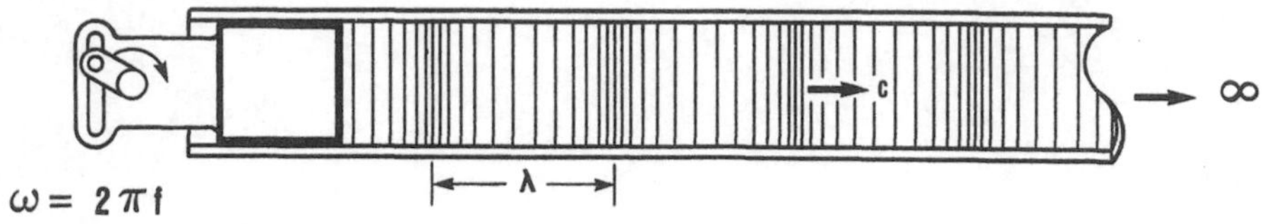

FIGURE 4.1 Plane wave generator.

When the piston begins to move, air in the tube is still at rest. Air is elastic and has inertia, so the elastic bonds between molecules in the thin layer at the piston face are compressed slightly as the piston motion is resisted by the inertia of these molecules. The pressure built up in this layer exerts itself on the next layer of molecules, and the pressure propagates down the tube at a rate referred to as the *speed of sound*. The air molecules themselves duplicate the piston motion and do not travel down the tube. The pressure waves that do travel down the tube occur when each molecule transfers momentum to the ones close to it.

As the speed increases, the pressure on the piston face also increases. This increased pressure continuously propagates down the tube so that points along the tube experience identical pressure-time signatures. However, since waves have finite speed, these pressure signatures occur at different times along the tube.

At midstroke, when the piston begins to slow down, the air is still moving at the piston's maximum speed. The pressure on the piston face begins to drop as the air momentum tends to pull the air away from the piston. When the piston reaches its farthest position, the pressure on its face has returned to the initial equilibrium pressure. As it begins to retract, pressure on its face is below equilibrium and is described as a *rarefaction*.

The pressure at the surface of the piston, at any instant is

$$p = \rho c \dot{x} \quad \text{Pa}$$

where

$\dot{x}$ = instantaneous piston velocity, m
p = air density, 1.18 kg/m^3
c = speed of sound, 344.4 m/sec
ρc = characteristic impedance, 406 N-sec/m^3

The values given above are for the normal room temperature of 22°C (or 72°F) and atmospheric pressure of 0.751 m (29.6 in.) of Hg.

If the piston motion is

$$x = r \sin \omega t \quad \text{m}$$

where

r = half the piston travel, m
ω = rotational speed, $2\pi f$, rad/sec

the piston velocity and the pressure are

$$\dot{x} = r\omega \cos \omega t \quad \text{m/sec}$$

$$p = \rho c r \omega \cos \omega t \quad \text{Pa}$$

The power transmitted to the air is equal to the average of the product of the piston force times its velocity. Since each term is a cosine function, their product is a cosine-squared function. As we saw in Section 3.2.1, the average of cosine squared is one-half the square of the cosine amplitude. The power, then, is

$$W = \frac{Ar^2 \rho c \omega^2}{2} \quad \text{W}$$

where A is the piston area in square meters.

These air pressure changes traveling down the tube are called sound waves because a plot of the pressure and time, at any point along the tube, looks like a wave. These waves are very weak, on the order of 0.0001 psi for a 90-dB sound. Although the equilibrium pressure is much greater, it is not part of the sound.

The term "sound wave" is not limited in use to single-frequency pressure variations, it applies as long as the pressure variation is periodic. Since pressure changes sinusoidally with time in our model, the sound is said to be a pure tone.

The sound wave travels down the tube at about 1100 ft/sec. This velocity changes slightly with temperature, going from 1087 ft/sec at 32°F to 1130 ft/sec at 68°F.

The distance between pressure maximums along the tube is the wavelength. This is a function of both frequency and the speed of sound

$$\text{wavelength, } \lambda = \frac{c}{f} \quad \text{m}$$

where

c = the speed of sound in air, approximately 1130 ft/sec or 344 m/sec
f = frequency, Hz

Later in this chapter we see how wavelength influences sound radiation. In other chapters we also see how it enters into room standing waves. Figure 4.2, which shows how wavelength varies with frequency, was prepared as a ready reference in these later studies.

When the diameter of the tube is small in comparison to the wavelength, the waves moving down the tube are called plane waves. This designation stems from the fact that the wavefronts, the surfaces having the same pressure at any given time, are essentially flat. The important feature of plane waves is that they pass through constant-cross-section paths and, except for friction losses, do not diminish with travel distance.

4.1.1 Wavefronts

Sound normally radiates spherically. In the model, sides of the tube prevent this type of distribution and the plane waves that resulted

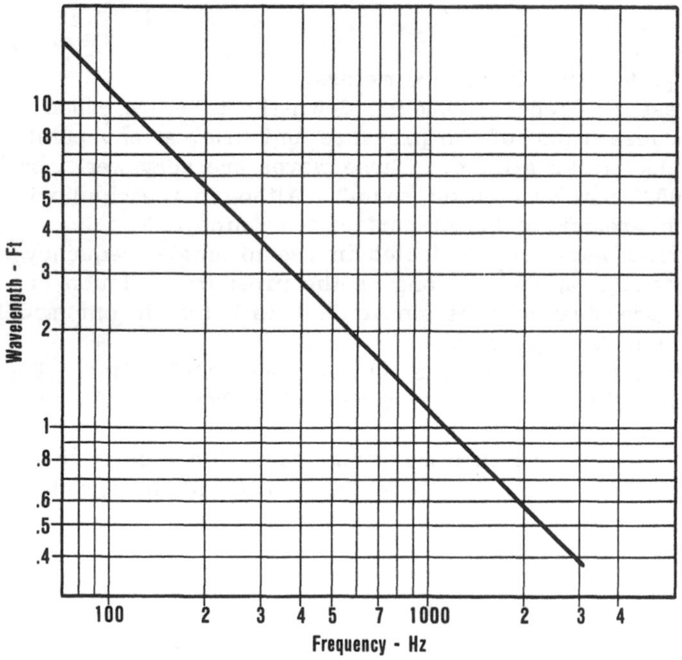

FIGURE 4.2 Wavelength in air.

are a special case. They are also the easiest to analyze. Wavefronts far from their source, although still spherical, have curvatures so small that they too are sometimes treated as planes to simplify analysis.

Sound from small sources is considered to radiate spherically from a point. Small, here, is relative to the wavelength of the sound. The shape and surface contour of the source is unimportant. Even a small hole in a wall, with sound passing through, is a point source.

Radiation from large sound sources have other-than-uniform spherical patterns, because of interference between the emissions from different parts of the surface. This effect is studied by dividing the source into sections and representing each with a point source having the same strength and phase relationships.

When sound waves, of the same frequency but from two different sources, meet at a point, they add as discussed in Section 3.2.2. The effect is as shown in Figure 4.3. This is a simple case of two point sources, a half-wavelength apart. These sources are taken to be in phase. The solid circles are spaced one wavelength apart and they show the locations where the waves are in phase with the sources. The smallest dashed circles have a half-wavelength radius and the rest of these circles are spaced one wavelength apart. They

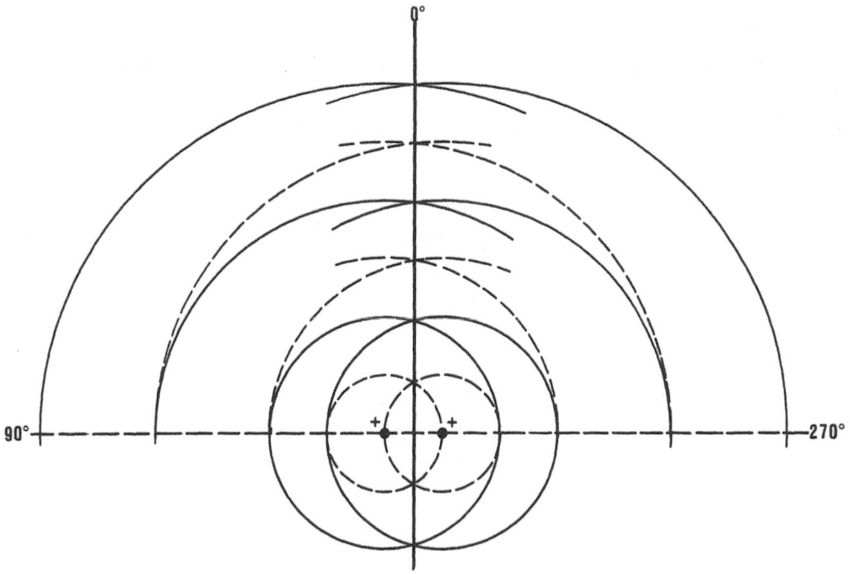

FIGURE 4.3 Effect of two point sources $\lambda/2$ apart.

all, therefore, are 180° out of phase with the sources and the solid circles. Wherever solid circles meet, the sound waves being in phase, add algebraically. Wherever a solid and a dashed circle meet, it indicates two waves 180° out of phase which cancel each other. The resulting pressure at these points is therefore zero.

The solid line at right angles to the axis through the sources is the locus of the points where maximum pressures occur. The dashed line through the sources indicates the locus of zero pressure points. Pressures at angles between these axes vary between zero and maximum as phase angles between the meeting waves vary from 180 to 0°.

Figure 4.4 shows the pressure distribution for this example and for cases where the sources are one-quarter and three-quarter wavelengths apart. The length of a radius vector from a point midway between the sources to the edges of the patterns indicates the strength of the combined wave in that direction, as measured at a distance from the sources. The patterns, of course, are three-dimensional and are symmetrical about the axis through the sources.

It can be seen from Figure 4.4 that the radiation is nearly spherical when the sources are close. As the separation increases, the radiation becomes less spherical, with the beams becoming narrower and additional beams being generated.

4.1.2 Reflection

Sound, like light, is reflected by changes in density. Unlike light, surface roughness has little effect on sound reflection because the irregularities are small in comparison to the sound wavelengths. A

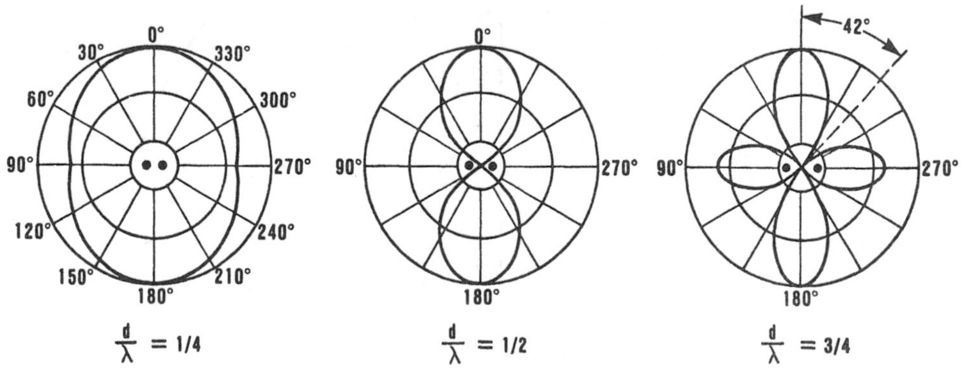

FIGURE 4.4 Radiation patterns around two point sources (dipoles). (From Ref. 2.)

solid concrete or brick wall will reflect almost all of the sound impinging on it. A more flexible surface that will allow some of the sound to pass through it will reflect the remainder of the sound. When sound impinges on porous surfaces, some penetrates the pores, where friction converts it into heat. Some may also pass through to the other side. The remainder is reflected.

When a noise source is located near a reflecting plane, sound reaches surrounding points by reflection as well as directly from the source, as shown in Figure 4.5. The two waves reaching a point add like any two waves having a phase difference. Phase difference occurs because the two paths are not equal in length. As a result, the sound pressure at various points surrounding the source is higher or lower than those that occur if the reflecting plane is not there.

The pressure distribution around the source is the same as if the reflective surface were replaced by a duplicate of the sound source, located as a mirror image as shown in Figure 4.5. This is referred to as an image source. The pressure distribution determined with this image technique, of course, is valid only on the real source side of the reflecting surface.

The case of reflections of wavefronts parallel to a surface is a special one. The sound pressure near the reflecting surface is the result of the approaching wave and the reflected wave

$$p = p_1 \sin\left(\omega t - \frac{2\pi x}{\lambda}\right) + p_2 \sin\left(\omega t + \frac{2\pi x}{\lambda}\right) \quad \text{Pa (rms)}$$

where

p_1 = approaching wave pressure amplitude, Pa (rms)
p_2 = reflected wave pressure amplitude, Pa (rms)
x = distance from reflecting plane, m
λ = wavelength, m

If the surface is rigid, the amplitudes of the two waves are essentially identical. We then apply a trigonometric identity from a handbook to get

$$p = 2p_1 \sin \omega t \cos \frac{2\pi x}{\lambda} \quad \text{Pa}$$

The sine term indicates that the pressure at any point continues to cycle at the same frequency.

The cosine term determines the amplitude of the pressure cycle at various distances from the reflecting surface. At the surface, where $x = 0$, the cosine term is 1. The amplitude is therefore $2p_1$ at the surface. This amplitude also occurs when $x = \lambda/2$ and at multiples

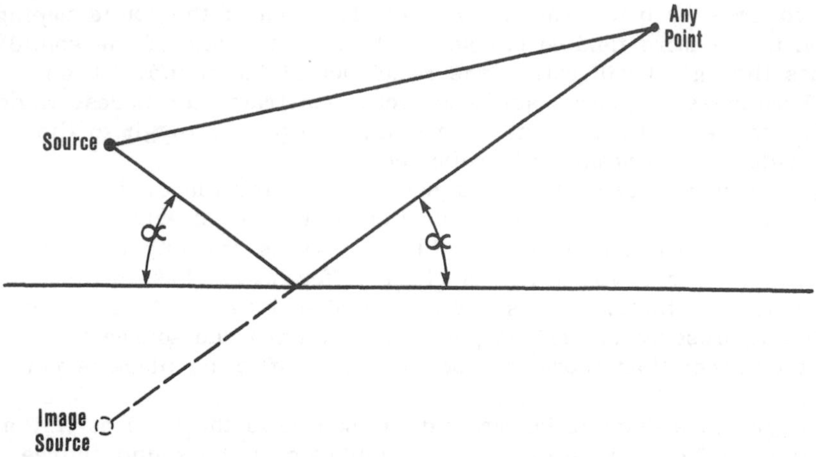

FIGURE 4.5 Sound reflections.

of this distance. At a distance of $x = \lambda/4$, the angle for the cosine term is $\pi/2$, so the cosine and the pressure amplitude are also zero. The amplitude is also zero at distances that are odd multiples of this distance.

4.1.3 Piston Sound Generators

Mathematical analyses of an oscillating piston surrounded by an infinite baffle provides insight into how radiator size affects radiation. Figure 4.6 shows how the radiation efficiency of pistons vary with the ratio of their circumference to the wavelength of the sound being radiated (3). Efficiency, here, is in terms of the amount of energy radiated from a unit of area, for a given amplitude. It can be seen that when the piston circumference is less than the wavelength, the efficiency is low. Analyses of other radiation models all show this same general characteristic, although the degree varies with the model.

Figure 4.7 shows how size affects the radiation pattern of a piston (4). The ratio used here is diameter to wavelength. A radius vector from the center of the piston face to the beam envelope indicates the beam strength in that direction. The patterns are three-dimensional and are symmetrical about the piston axis. Small pistons tend to radiate equally in all directions. Directionality increases with size and the beam becomes sharply focused when the piston is relatively large in comparison to the wavelength. With a given piston size, beam sharpness increases with increasing frequency.

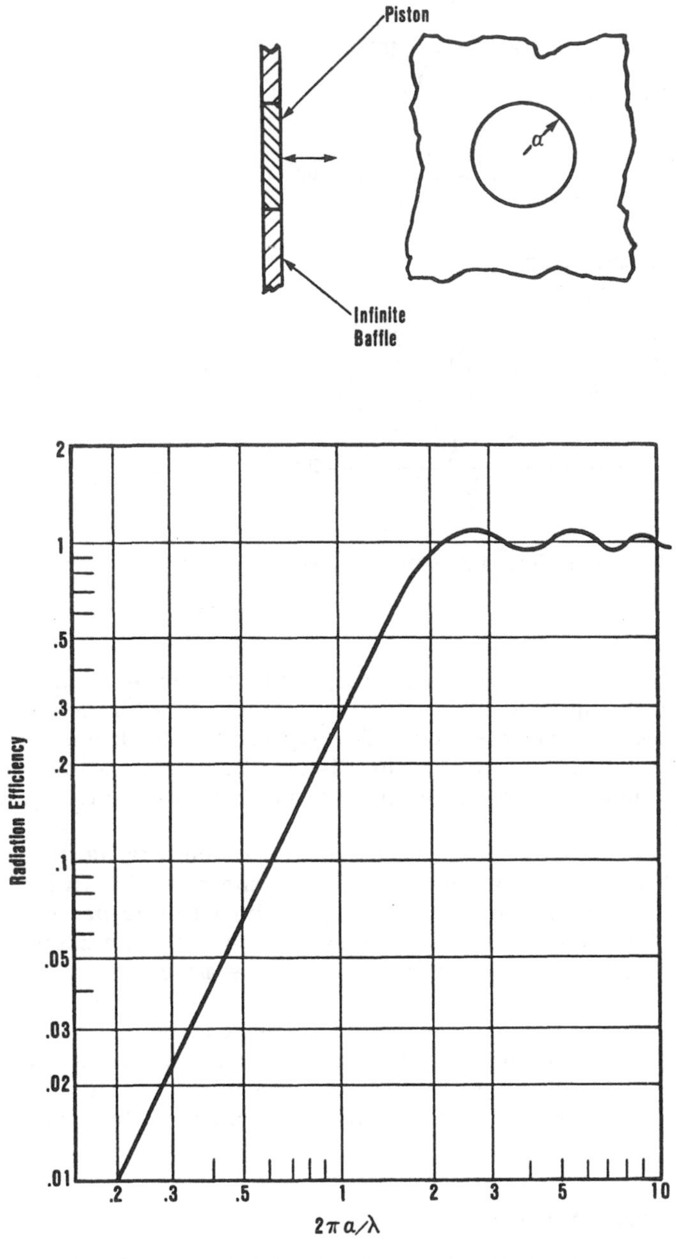

FIGURE 4.6 Baffled piston radiation efficiency. (From Ref. 3.) Reprinted with permission from Acoustical Society of America, © 1986, Acoustical Society of America.

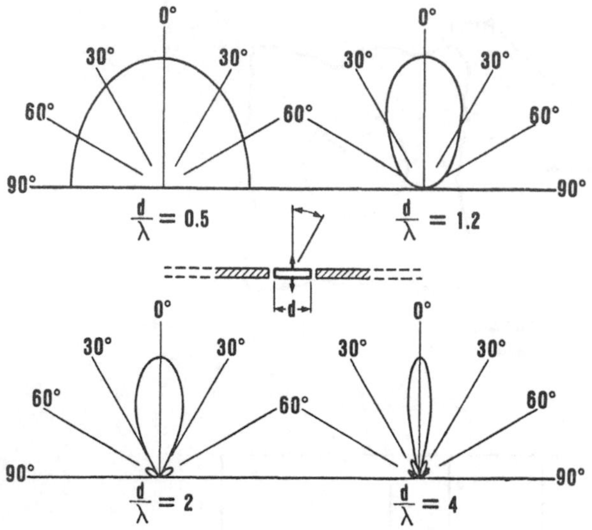

FIGURE 4.7 Piston radiation patterns. (From Ref. 4.)

Figure 4.8 is similar to Figure 4.6 except that the scale of the abscissa has been changed to illustrate the relationship with frequency. A 6-in. piston diameter was assumed. For reference, the frequency of sound with a wavelength equal to the piston circumference is 720 Hz. This curve moves to the left as piston size increases and to the right as size decreases.

The radiation efficiencies of two other piston models are plotted in Figure 4.8. These are a piston operating in the end of a long tube and a piston operating in the open (5). The latter is particularly inefficient because at low frequencies, air from the pressure side circulates around the piston and cancels out some of the rarefaction being generated on the other side.

At frequencies below 800 Hz, the piston in the end of a tube is twice as efficient as the piston in a baffle. Like the baffle, the tube prevents the circulation that occurs with the free piston. The increase in efficiency results from the piston in the tube radiating from both faces.

4.1.4 Multipole Sound Sources

Earlier discussions involving point sources all considered sources in phase with each other. Sources can have any phase relationship with adjacent sources. However, in almost all practical problems adjacent

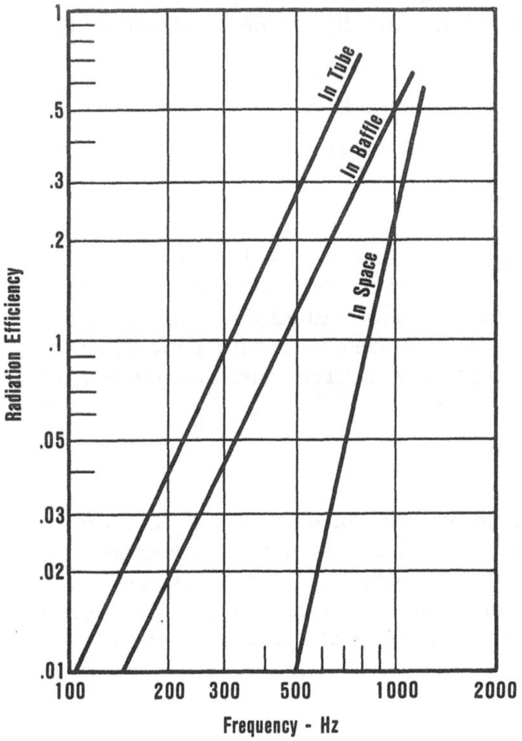

FIGURE 4.8 Six-inch piston radiation.

sources are parts of the same vibration system and are either in or 180° out of phase. Here we discuss the later case.

The models used to study the effects of adjacent sources consist of arrays of pulsing spheres. Alone, these radiate spherical wavefronts and are called *monopoles*. When two are close and are out of phase, considerable canceling takes place. This effect has been described as "sloshing" of the air adjacent to the sources and it greatly reduces the net power radiated by the two sources. Pairs of this type are referred to as *dipoles*.

In the earlier discussions, the piston in a baffle is a monopole. Similarly, the free piston is a dipole.

If two dipoles are close to each other and oriented so that opposite poles are close, considerably more cancellation or sloshing occurs. This combination is referred to as a *quadrupole* and is far less efficient in radiating sound than even a dipole.

Figure 4.9 compares the radiation characteristics of the various types of sources. Symbols used in this figure are defined as follows:

A = source area
v = root-mean-square source velocity
k = wave number, k = $2\pi/\lambda$
c = speed of sound in air
ρ = density of air

It can be seen that the close presence of another out-of-phase sound source greatly reduces radiation. The addition of another dipole to form a quadrupole squares the reduction. This effect diminishes as the source separation increases. With separations greater than a quarter-wavelength the sources are independent of each other.

4.1.5 Tube Radiation

Hydraulic lines are a common source of noise, so their radiation characteristics are of great interest. The noise that they radiate comes from two sources. The first is fluidborne noise that causes the line to pulse with pressure, producing monopole radiation. The second is structureborne noise which causes lateral bending vibration and

Source Type		Radiated Power
1. Monopole or simple source	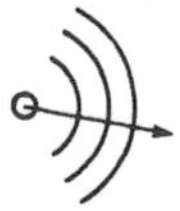	$W_{rad} = W_o = \frac{\rho c k^2}{4} A^2 (\upsilon^2)$
2. Dipole source	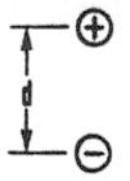 kd <1	$W_{rad} = 1/3(kd)^2 W_o$
3. Quadrupole source	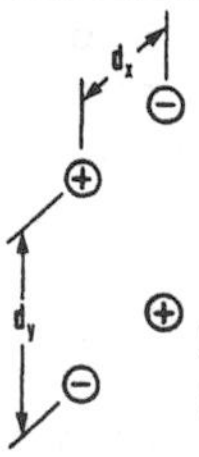 kd_x <1 kd_y <1	$W_{rad} = 1/15(kd_x)^2(kd_y)^2 W_o$

FIGURE 4.9 Multipole sources radiation. (From Ref. 6.)

dipole radiation. Each unit line length acts as a point source in phase with its neighbors. The combined effect is that sound radiates in cylindrical rather than spherical waves.

Figure 4.10 compares the radiation efficiency of a 2-in. line in the two modes. At low frequencies the vibrating line is a very poor radiator in comparison with the pulsating line. It is subject to front-to-back circulation like the free piston, while the pulsating line is free of this effect. Despite the great difference in their radiation efficiencies, both modes are important noise sources and are discussed in more detail in Section 9.2.2.

The sound radiated from tubes excited by inside fluidborne energy has been studied extensively (7–10). Most of these studies focus on large piping used in process industries. These have more complex vibration patterns than those of hydraulic piping. Only Hope (10) deals exclusively with hydraulic lines. He also examines radiation from cylindrical hydraulic components and line vibration caused by fluidborne noise.

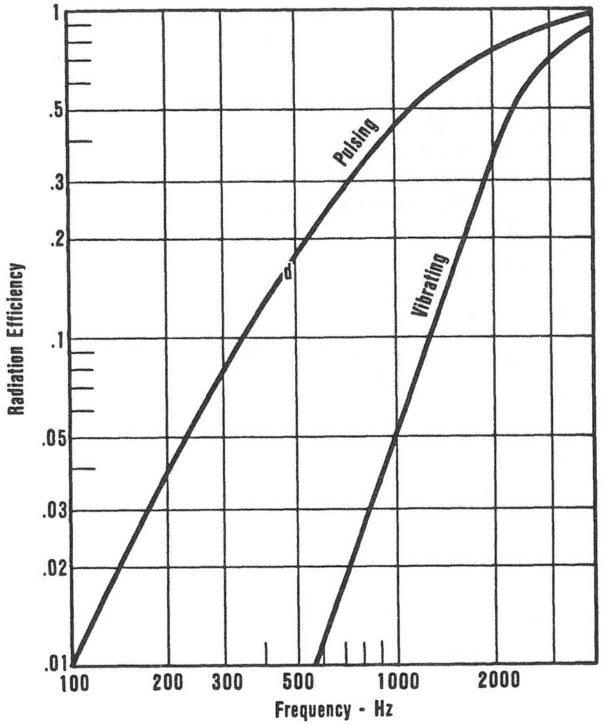

FIGURE 4.10 Two-inch hydraulic line radiation.

4.1.6 Plate Radiation

Plates or sheet metal panels are common sources of multipole radiation. As discussed in Section 3.1.4, these members vibrate in many modes. In the higher modes they have a large number of areas of strong vibration, as shown in Figure 4.11. Each vibration area is outlined by node lines free of vibration and is out of phase with those adjacent to it.

In the central area of the plate, portions of the vibrating areas can be grouped into quadrupoles, as shown. Similarly, areas around the edges can be grouped into dipoles. Only the corners are not subject to multipole cancellation from adjacent areas, and so most of the radiation from these plates comes from their corners.

At low frequencies, where the dimensions of the plate are less than the wavelength, the corners also interact with each other. In this case they behave like dipoles or quadrupoles, depending on the phase relationships that exist. Small plates therefore tend to be very poor radiators at low frequencies.

At higher frequencies, *edge modes* like the one shown in Figure 4.12 can occur. These are more efficient radiators since the long edges act as monopoles.

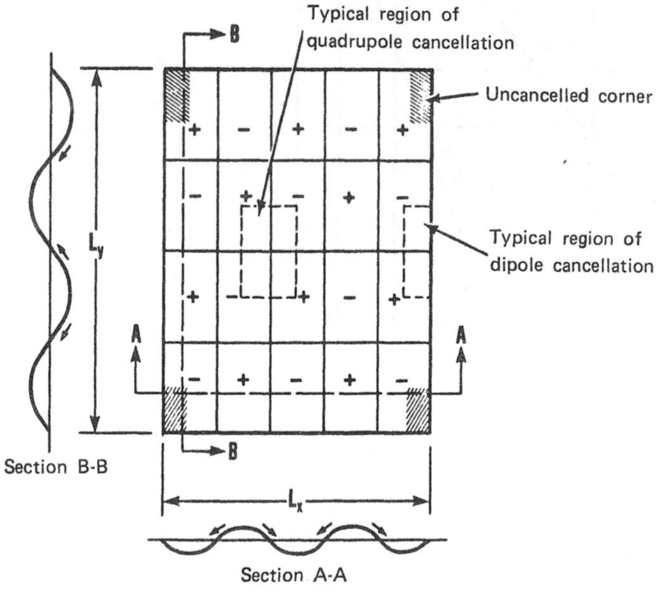

FIGURE 4.11 Plate corner radiation mode. (From Ref. 11.)

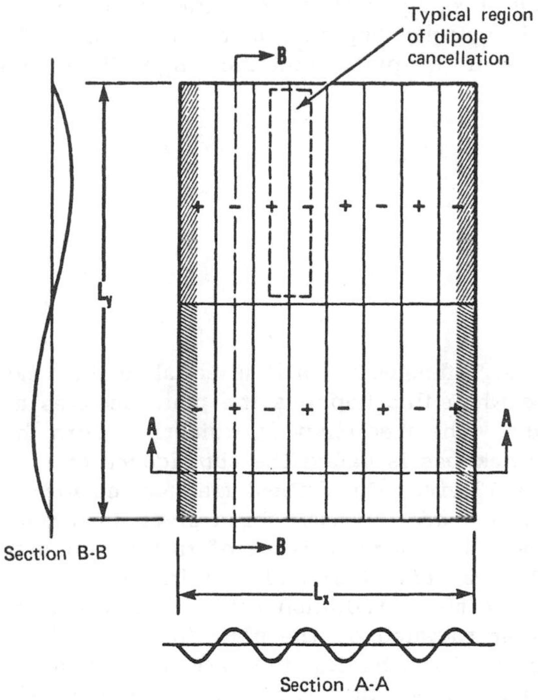

FIGURE 4.12 Plate edge radiation mode. (From Ref. 12.)

Stress waves travel through plates in the same way that sound waves travel through air. They travel at the *longitudinal wave velocity*, which is a function of the plate material. This velocity is

$$c_L = 71.5\sqrt{\frac{E}{w}} \qquad \text{ft/sec}$$

where

E = modulus of elasticity, psi
w = weight per cubic foot, lb/ft

The constant primarily takes care of the mixing of feet and inches in the variables and for converting weight into density. It also includes a small plate rigidity correction involving Poisson's ratio.

Lateral bending deflections propagate through plates at the *plate bending wave velocity*. This velocity depends on the frequency of the deflections, the thickness of the plate, and the longitudinal wave velocity

$$c_B = \sqrt{0.15tfc_L} \qquad \text{ft/sec}$$

where

t = plate thickness, in.
f = frequency, Hz

This velocity has special significance when it is equal to the velocity of sound in air because when this happens the plate becomes a very efficient sound radiator. The frequency at which it occurs for a given plate material and thickness is called the *coincidence* or *critical frequency*. Figure 4.13 gives the critical frequencies for steel and aluminum plates, which are the same for a given thickness.

Figure 4.14 shows the radiation characteristics of simply supported 3- by 4-ft plates that are 1, 1/2, 1/4, 1/8, and 1/16 in. thick. Radiation in this figure is in terms of radiation ratio (13), which is the ratio of the acoustic power radiated by the plate to that radiated by an equal area of an infinite piston having the same velocity as the plate. The plate velocity is the space-time rms value. Sections of these curves due to corner and edge radiation are indicated.

At the critical frequency the plate deflection wavelength and the sound wavelength are equal. This is similar to resonance and causes the peaks in the curves.

4.1.7 Near and Far Fields

In these radiation discussions, the patterns resulting from various influences exist only at a distance from the radiation surface. In the case of a panel with much of its area consisting of quadrupoles, the surface sound pressures are high at the centers of the vibrating elements. The interactions between elements of a quadrupole occur in the *near field*, which extends out from the radiating surface. Once the additions and cancellations of the waves from the various point sources are accomplished and the beams are established, the radiation becomes spherical. The space where the radiation is spherical is called the *far field*. The presence of spherical radiation is confirmed by a series of measurements, made along a radius from the center of the sound source, which indicates where pressures become inversely proportional to the distance from the source.

The demarcation between the near and far fields depends on the nature of the point sources and their distribution. An often used

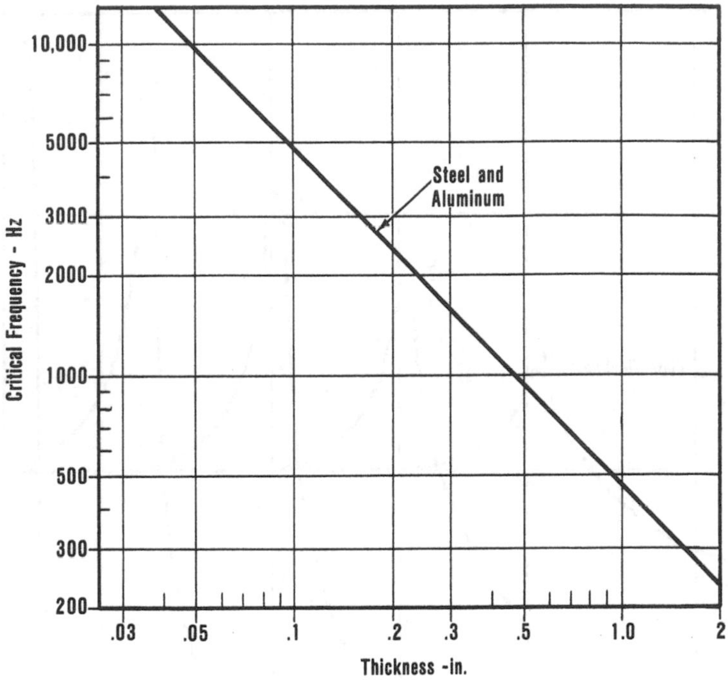

FIGURE 4.13 Plate critical frequencies.

rule of thumb is that the near field extends out a half-wavelength from the surface. Sound measurement standards specify minimum measuring distances in terms of source size, suggesting that the extent of the near field is related to the size of the radiator.

4.1.8 Aperiodic Sound

The sound waves that have been discussed are classed as periodic because even though they may have long, complex time profiles, the sum of many frequencies, they repeat periodically. Some hydraulic noises, particularly those caused by cavitation, are random in nature. They do not repeat and are classed as aperiodic. They are hissing sounds and are referred to as *broadband noise*.

Usually, these sounds have fairly constant average levels. When one of these sounds has the same level over a wide frequency range, it is called *white noise*. For analysis, broadband noise is treated as though it consists of a series of frequencies that are on the order of 1 Hz apart.

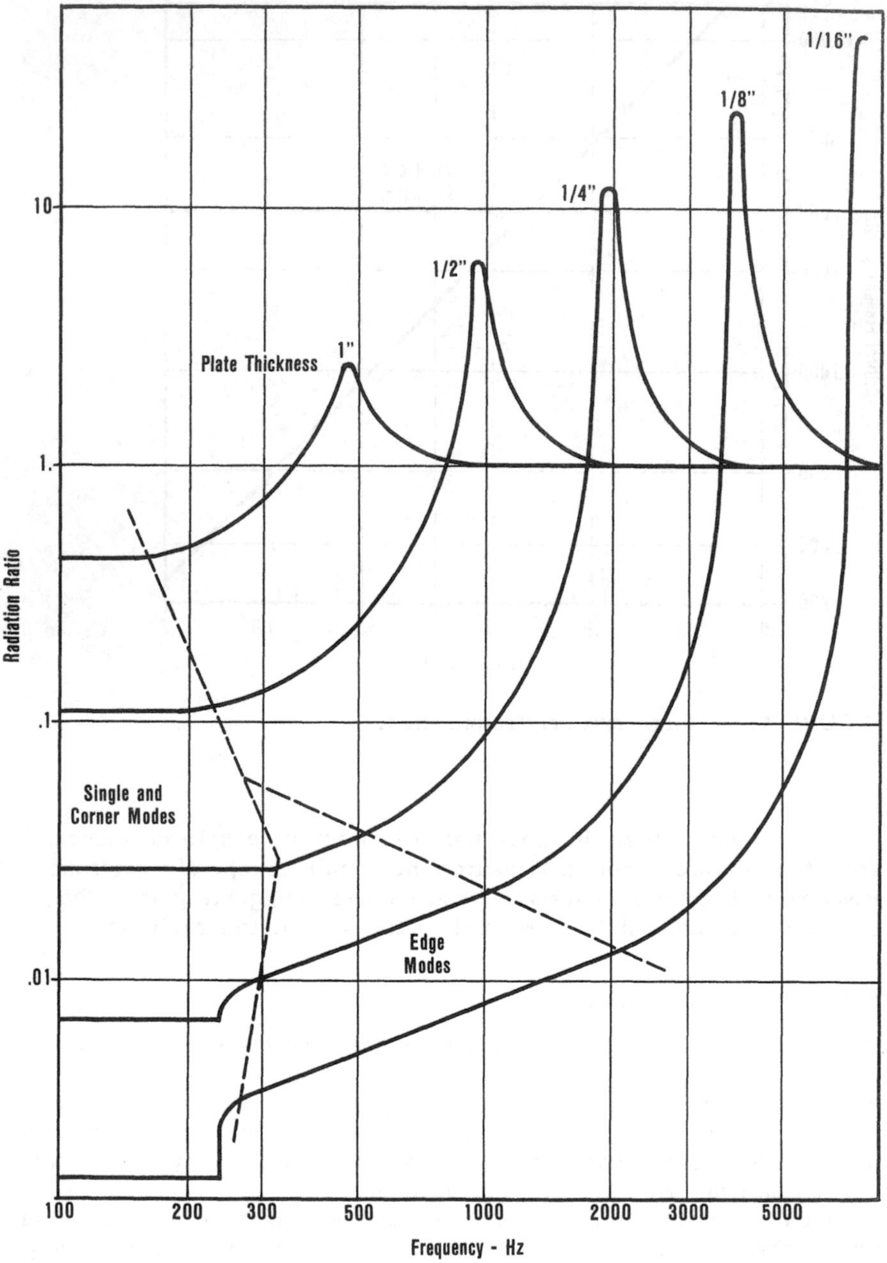

FIGURE 4.14 Plate radiation characteristics.

4.2 SOUND STRENGTH

The strength of a sound is expressed in terms of power, intensity, or pressure.

4.2.1 Power and Intensity

Strengths of combinations of sounds are determined by adding the sound power of the individual sounds. This applies when multiple noise sources in a machine contribute to the total noise of the machine or a Fourier series of sound components combine into a complex sound. The mean-square power of such combined sounds is the total of the mean-square powers of all its component frequencies

$$\text{total sound power } W_T = W_1 + W_2 + W_3 + \cdots + W_n \quad \text{W}$$

where

1, 2, n = number of sounds or sound components
W_1, W_2, W_n = powers of contributors, W

In the piston and tube example of sound generation, presented in Section 4.1, sound power is transmitted down the tube by wave action. There is very little resistance to dissipate energy, so the power will remain the same in passing through parallel planes along the tube. This is also true of the power passing through concentric cylindrical surfaces or through concentric spherical surfaces. In the latter two cases, however, the imaginary surfaces receiving the power increase in area as the distance from the source increases. This means that the power reaching a unit area of these surfaces, the *intensity* (I) decreases in inverse proportion to these areas.

In the case of spherical waves, as the radii of the spherical surfaces increase, their area increases as the square of these radii. So the power intensity decreases at the inverse of this rate

$$\text{sound intensity } I = \frac{W}{4\pi r^2} \quad \text{W/m}^2$$

where

W = sound power, W
r = radius of surface, m

The cylindrical waves that radiate from hydraulic lines increase lineraly with radius. The comparable equation for these waves is

$$I = \frac{W}{2\pi r} \qquad W/m^2$$

Sound power intensity is a vector quantity having both magnitude and direction. In the last decade or so, instrumentation for measuring sound intensity has become readily available. It measures the sound power flow in the direction of an axis through two microphones that are held a fixed distance apart. When intensity is measured, integration of its magnitude over the entire area of an imaginary surface surrounding a noise source determines the total sound power radiated by the source. This instrument, which includes a minicomputer, is quite expensive compared to other types of sound-measuring equipment, so it is not commonly used.

Since sound intensities are related to power, their mean-square values, like those of power, also add algebraically. However, care must be taken to add just those having the same direction.

4.2.2 Pressure

Generally, sound intensity is determined from pressure measurements which require only relatively simple instruments. Conversion is made from

$$\text{sound intensity } I = \frac{P_{rms}^2}{\rho c} \qquad W/m^2$$

where

P_{rms} = root-mean-square sound pressure, PA
ρc = characteristic resistance of air, 406 N sec/m^3

In practice, this equation is only of academic interest because sound measurement units were selected so that sound pressures and sound intensities are numerically equal.

The equation above holds in practice only when the sound comes from one direction. Sound pressure is a scalar quantity and is not effected by direction, whereas intensity, a vector, is dependent on direction. In measurements made with the sound coming from multiple sources and directions, the pressure is the sum of the emissions from all of them. There is an intensity for each source equal to the pressure contributed by that source, but these must be added vectorially.

It should be noted that pressure is measured in terms of *root mean square*. The unit is the pascal, which is a relatively new term for the newton/square meter and dyne/square centimeter used in older acoustics books.

When we want to determine the sound pressure of a complex sound from its component pressures or the total sound pressure of several sounds, we take the square root of the sum of the squares of the pressures of the individual contributors

$$\text{total sound pressure } P_T = \sqrt{P_1^2 + P_2^2 + P_3^2 + \cdots + P_n^2} \quad \text{Pa (rms)}$$

where

1, 2, n = number of sounds or components
P_1, P_2, P_n = pressures of contributors, Pa (rms)

4.3 DECIBELS

Sound pressures vary over a range of over five orders in magnitude (10^5) and sound powers over twice that number. If we tried to cover such ranges with ordinary numbers, it would be very unwieldy.

4.3.1 Logarithms

For this reason a logarithmic system of measurement is used. This causes considerable confusion for people unfamiliar with logarithms. However, after some use, they are found to provide many advantages.

Logarithms to the base 10 are used for the scales in this system and the unit is the decibel. Although it is not obvious, the bel is named for Alexander Graham Bell, the inventor of the telephone (14). For this reason the B in the abbreviation is always capitalized.

Whenever the decibel scales are used, sound quantities have the word *level* added to their names. We then speak of sound *pressure level*, sound *intensity level*, and sound *power level*. The equations defining these levels are as follows:

$$\text{sound power level } L_W = 10 \log \frac{W}{W_0} \quad \text{dB}$$

where

W = sound power, W
W_0 = reference sound power, 10^{-12} W

$$\text{sound intensity level } L_I = 10 \log \frac{I}{I_0} \quad \text{dB}$$

where

I = power intensity (from one direction through unit area), W/m^2
I_0 = reference sound intensity, 10^{-12} W/m^2

$$\text{sound pressure level } L_P = 10 \log \frac{p^2}{\bar{p}_0^2} \quad \text{dB}$$

$$= 20 \log \frac{p}{p_0}$$

where

p = rms sound pressure, Pa
p_0 = rms reference pressure, 20 μPa

4.3.2 Comparing Sound Levels

When a logarithm of a number is subtracted from that of another number, the result is the logarithm of the ratio of the second number to the first. This fortunate property makes comparisons of sound levels, in decibels, quite easy. The process is simply subtracting one level from the other and finding the antilog of their ratio from the table in Appendix B or from the following equations:

$$\text{sound power ratio} = 10^{\Delta \text{dB}/10}$$

$$\text{sound pressure ratio} = 10^{\Delta \text{dB}/20}$$

where ΔdB is the difference in decibels.

Most people working in noise control soon learn a few conversions and references to the tables become infrequent. One reason for this is that it is seldom necessary to know the ratios between sound levels precisely. The other is that for each 10 dB that is subtracted from the difference, the power ratio is multiplied by ten. It takes a subtraction of 20 dB for the pressure ratio to be multiplied by 10. After such subtractions the remainder will easily be compared, mentally, with the few reference decibel levels given in Table 4.1.

An example can be drawn from this table, but first it must be noted that the ratios for power are equal to the square of those for pressure, for a given decibel difference. Assume that we have a difference in two sounds of 32 dB and we want their pressure ratio. Subtracting 20 leaves a balance of 12, which is a ratio of 4. The

TABLE 4.1 Short Table of Decibel Conversions

Sound-level differences (dB)	Sound pressure ratio	Sound power ratio
1	1.1:1	1.3:1
3		2
6	2	4
10	3.2	10
12	4	
20	10	10^2

Source: Appendix B.

20-dB subtraction multiplies this by 10, so the ratio of the two sound pressures is therefore the product of 4 × 10 = 40.

It should also be noted that 12 dB could have been considered as two 6's, each having a ratio equivalent of 2. The ratio would then be computed as 2 × 2 × 10 = 40.

If in this example we wanted the ratio of sound power levels, we would first have subtracted 30, which corresponds to a factor of 1000. The remainder, 2, is not included in Table 4.1. Although its equivalent can be found in the table given in Appendix B or by calculation, it generally is enough to know that the ratio of the two powers is between 1300 and 2000.

4.3.3 Adding Sound Levels

When sounds are combined, their total power is the arithmetic sum of the powers of the individual sounds. This is the basis for combining all sounds, as discussed in Section 4.2.1. Regardless of whether we are adding power, intensity, or pressure decibels, the same process is used.

Sound levels can be added by using Figure 4.15. To add two sounds we take the difference in their levels and from this chart find the amount that must be added to the highest to get their total level.

When more than two sounds or sound components are to be added, the two highest are first added together. The total of these two is then treated like one sound and the next highest is added to it. This process is repeated until all the sounds have been accounted for or until the difference between a subtotal and the next level to

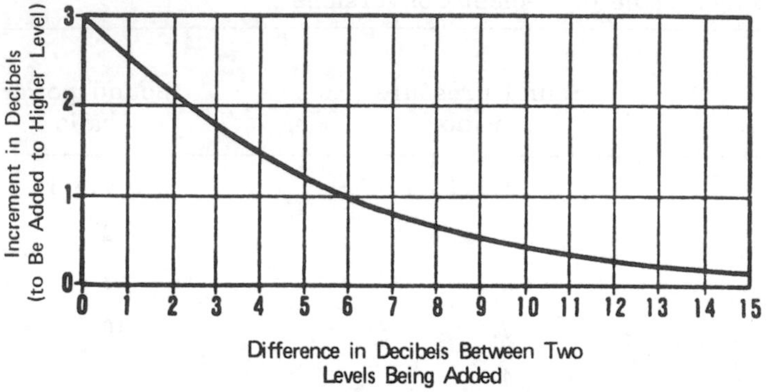

FIGURE 4.15 Adding sound levels. (From Ref. 15.)

be added is so great that additional corrections are insignificant. The normal variability in sound measurements makes it impractical to calculate levels to tenths of a decibel.

For an example of the use of Figure 4.15, assume that pressure levels of four noise components of different frequencies are to be added. These sound pressure levels are 84, 87, 78, and 86 dB. The two highest, 87 and 86, are added together first. Their difference is 1 dB and the chart indicates that 2.5 is added to the 87 to get a subtotal of 89.5. The 84-dB level is now added to the 89.5: Since their difference is 5.5, from the chart, this adds 1.1 dB and the subtotal becomes 90.6 dB. Since the 78-dB component is 12.5 dB below this subtotal, it adds only 0.2 dB to the total. With rounding, the total of the four components is therefore 91 dB.

Where a computer or scientific calculator is readily available, it may be easier to use them to add sound levels. The total sound level is calculated from

$$L = 10 \log\left(10^{L_1/10} + 10^{L_2/10} + 10^{L_3/10} + \cdots + 10^{L_n/10}\right)$$

where

$1, 2, n$ = numbers of sound levels to be added
L_1, L_2, L_n = sound levels to be added, dB

These methods are not valid for adding sounds of the same frequency, as discussed in Section 3.2.2. However, in the case of adding hydraulic moises which have harmonic frequencies in common, the errors incurred generally appear to be acceptably small. This may be due to very small differences in these frequencies, causing beat notes that extenuate the errors or because large differences in the levels at the common frequencies are so great that only small errors can occur. If errors due to adding equal frequencies do occur, the calculated levels will be higher than the actual.

4.4 SOUND MEASUREMENTS

Sound is measured with a microphone which generates an electrical signal that is an analog of the sound pressure. Although this discussion is directed toward airborne noise measurement, it is equally applicable to the measurement of fluid and structureborne noise. Transducers used to sense these noises also produce their electrical analogs. The same instrumentation is used to measure and analyze these noises.

The simplest noise measurement is one in which the electrical signal is analyzed with a sound-level meter. This meter indicates the root-mean-square average of the signal in terms of decibels. In the case of airborne noise, the resulting statistic is the *total sound pressure level*. Here, "total" refers to the fact that all the sound, regardless of frequency, is measured.

Data of this type do not tell anything about the frequency distribution of the sound. To increase the information derived from such measurements, electrical filters are generally imposed between the microphone and the meter. These filters attenuate or eliminate certain frequencies before they reach the sound-level meter.

4.4.1 Noise Frequency Data

Some simple filters are used to measure total sound but give greater weight to certain frequencies. These are particularly useful in providing single-number sound-pressure-level ratings such as those found in government regulations and in purchase specifications. A-weighted or dB(A) levels, mentioned in Chapter 1 and later in Section 4.4.3, are of this type. The basis for these weightings is covered in Section 4.5.

Filters that can be adjusted to any acoustic frequency and pass only a narrow band about the tuned frequency are called *discrete frequency filters*. When these are swept through the range of sound frequencies they identify the frequencies, as well as the strengths, of the single-frequency harmonics of a noise.

This type of data is very valuable in research and in tracking down unknown noise sources. Sometimes the volume of this data leads to handling problems and is often more information than needed. For this reason most noise frequency distributions are measured with a series of fixed-frequency, wider-band filters that span the appropriate range of frequencies.

Commonly, *octave band* filters are used. They are given this name because their upper frequency limit is twice their lower limit. These filters are standardized (ANSI S1.11) and 10 of them cover the range from 20 to 20,000 Hz, as seen in Table 4.2. The two lowest and the two highest are seldom used in hydraulic noise work. With pumps and motors we know their fundamental and harmonic frequencies, so the sound levels in the middle six octaves is all that we need to know about a noise.

Pump noise is generally rated at many combinations of speeds, pressures, and displacement settings. Data handling in these evaluations is, therefore, a big task. Since octave band analysis provides the minimum data required for ratings, it is favored for such work.

Third-octave filters are used where better definition of pressure distributions is desired. These are a compromise between discrete frequency and octave band measurements. They single out and individually measure more noise harmonics than octave bands but have the disadvantage of having to handle three times more data.

4.4.2 Sound Power Measurements

The accepted way to deal with small sound sources, such as pumps, motors, and valves, is to measure the total sound power that they emit. This is an intrinsic parameter that does not depend on external conditions such as the acoustic characteristics of the test space. With large machines this tends to be too difficult, so it is customary to settle for simpler sound pressure measurements.

Sound power can be measured accurately in a variety of test spaces as long as extraneous sounds do not interfere. Many test facilities attempt to provide an anechoic or acoustic free field in which sound radiates without some of it being reflected back towards the source. This is done by making the walls highly absorptive. The advantage of such space is that only sound coming from the source is measured and sound pressures decrease 6 dB each time the distance from the source is doubled, so differences in measuring distances are correctable.

It is difficult to operate in an anechoic test room because the floor is covered with absorptive material. Therefore, a compromise is often made and a semianechoic field is provided. This is one that has a reflecting plane near the test unit. As discussed in Section 4.1.2, radiation with the reflector present is the same as if there were a mirror image of the test unit on the other side of the plane. The acoustic center of the sound source and its virtual image is in the plane at the center of their projections. When the source and reflector

TABLE 4.2 Filter Band Frequencies

Limit frequency	1/3-Octave band center frequency	Octave band center frequency
22		—
	25	
28		
	31.5	31.5
36		
	40	
45		—
	50	
56		
	63	63
71		
	80	
89		—
	100	
112		
	125	125
141		
	160	
178		—
	200	
224		
	250	250
282		
	315	
335		—
	400	
447		
	500	500
562		
	630	
708		—
	800	
891		
	1,000	1,000
1,122		
	1,250	
1,413		—
	1,600	
1,778		

TABLE 4.2 (Continued)

Limit frequency	1/3-Octave band center frequency	Octave band center frequency
	2,000	2,000
2,239		
	2,500	
2,818		—
	3,150	
3,548		
	4,000	4,000
4,467		
	5,000	
5,623		—
	6,300	
7,079		
	8,000	8,000
8,913		
	10,000	
11,220		—
	12,500	
14,130		
	16,000	16,000
17,780		
	20,000	
22,390		—

are close, the mirror image just doubles the sound pressures directly radiated from the test unit. Otherwise, the sound field behaves like an anechoic field, with sound radiating from the acoustic center.

Sound power is determined by integrating sound intensity over an imaginary hemisphere about the acoustic center. Since all of the sound comes from one point, intensities are numerically equal to measured sound pressures. The integration is accomplished by multiplying the average pressure level, at a given distance from the acoustic center, by the area of a hemisphere with a radius equal to this distance. Both the American National Standards Institute (ANSI) and the International Standards Organization (ISO) have standards covering these procedures and their technical requirements (ANSI S1.34 and ISO 3744). These standards are practically identical. This is the case of all standards with the same subject promulgated by these two organizations.

Even semianechoic rooms are difficult to use in testing pumps and motors. Drive shafts must be enclosed to keep their noise radiation from adding to that of the test unit, and this interferes with the free field. This problem is solved by adding another reflecting plane. The test unit, then, is at the junction of two perpendicular reflectors, and measurements are made over a quadraspheric surface.

This scheme works as well as the hemispherical measurement system. However, neither ANSI nor ISO have standards to guide such measurements. The British Standards Institution (BSI), seeing the need, has developed one (BSI 5944, Part 5).

The other extreme in measuring spaces is one in which the surrounding walls are very reflective. Reflections from them create a reverberant field that has the same pressure level throughout the room, except for a small space surrounding the sound source. This level is a function of the room size and reflectivity as well as the sound power being radiated. The relationship between sound pressure and sound power in these rooms is determined by calibrating the room with a sound source that has a known sound power output.

The advantage of such rooms is that sound power is determined from only one measurement. This was a big advantage before multiplexing and computer-averaging instrumentation became available. Since the same level exists throughout most of the room, the microphone can be located almost anywhere.

Reverberant rooms work only as well as just described when the sound is random or broadband, like that produced by valves. Pumps and motors, however, produce discrete frequencies, and these produce room resonances that cause sound pressure levels to vary throughout the room. To overcome this difficulty, rooms are equipped with moving vanes or diffusers which have the effect of moving the standing waves around. A stationary microphone, then receives a range of sound levels whose time average is equal to the room's average level. The same effect is achieved by moving the microphone about the room.

Both ANSI and ISO have standards for making measurements in such rooms (ANSI S1.33 and ISO 3743). Both of these standards address the problems caused by discrete frequency sound. These standards are for engineering grade measurements and refer to the test space as *special reverberation test rooms*. There are other standards for making more precise measurements in reverberant rooms, but these are not needed for hydraulic components.

4.4.3 Hydraulic Component Rating

The noise of hydraulic components is measured in terms of octave band and A-weighted sound *power* levels. This is done in any of the acoustic test spaces, using the appropriate standards discussed above. However, they are rated in terms of sound *pressure* levels. These sound pressure levels are arrived at by considering the sound to be radiated by a point source in a semianechoic space. The rating is the A-weighted sound pressure level 1 m from the point source (NFPA T3.9.12; NFPA T3.9.14; ISO 4412, Parts 1 and 2; BSI 5944, Parts 1 and 2). In practical terms, the rating is derived by subtracting 8 dB from the measured A-weighted sound power level. As discussed in Section 12.1.2, this rating is the noise level expected from machines using the subject unit.

All pump and motor noise standards require that only the sound radiated from the test unit be measured. This is difficult and a whole technology has evolved for excluding unwanted noises from support brackets, bed plates, drive shafts, and hydraulic lines. Some of the techniques are discussed in Section 9.3.5. Much of the development occurred in Great Britain and has been made into an information part of the British hydraulic equipment noise standard (BSI 5944, Part 3).

Valves present a different problem. Their noise is due to a cavitation plume streaming downstream from the valve (16). For this reason the sound radiated by the first few feet of line as well as that of the valve must be included in the measurement.

Great Britain has the only noise standard for hydraulic fluid power valves (BSI 5944, Part 4). The fluid power qualification is used here because there is a considerable amount of literature on the noise of process industry valves, and there may soon be a standard for such units. Since these valves are so different, it is doubtful if this standard would be applicable to fluid power valves.

The British standard requires that the test valve discharge line be flexible hose, coiled around the valve in about a 1-m-diameter circle. This is done to keep the total noise source compact and easier to measure. The hose is connected to the valve with a right angle elbow to reduce the probability of the cavitation plume extending beyond the 10 ft of hose wrapped around the valve. This standard is applicable in both semianechoic or special reverberant tests.

4.4.4 Machine Noise Measurements

Machines radiate noise from many locations and therefore are not characterized by a point source. Since the primary interest in machine noise is its relationship to OSHA limits, this is satisfied by measured sound pressure levels.

Machine noise sound pressure levels are measured at ear height at the operator's position and at intervals around the machine perimeter (NMTBA, ANSI S1.13). If the operator is seated, ear height is taken to be 1.1 m above the floor; standing and around the perimeter it is 1.5 m. Perimeter measurements are made 1 m from the machine boundary. The boundary, for this purpose, is often defined as the retangular parallelpiped that just encloses the machine if minor machine projections are disregarded. The boundary can also be a cylinder or an irregular shape as long as it is generally 1 m from the principal machine surfaces.

4.5 SUBJECTIVE MEASUREMENTS

To this point, sound has been viewed as strictly a physical phenomenon. As discussed in Chapter 1, our objective in controlling noise is to make it physiologically acceptable. To do this effectively, it is necessary to know how sound waves affect human beings. Only a minimal understanding will be attempted. If greater depth in this fascinating subject is desired, it is available in Refs. 17–19.

Usually, noise reduction is related to hearing conservation. However, there is some controversy regarding what levels are critical. Many studies of the relationship between sound levels and hearing loss have been made and a great deal is known, but not as precisely or convincingly as desired. One conclusion that appears to be drawn from these studies is that the damage is related more to perceived loudness than simply sound levels.

4.5.1 Loudness

Ear accuity varies with frequency as shown in Figure 4.16. These data were developed in 1956 by Robinson and Dadson at the National Physical Laboratory in Tedington, England, and have been accepted as an international standard (ISO R226). The curves indicate the sound levels that are needed to produce equally loud pure tone sounds at various frequencies. It can be seen that 4000-Hz tones are the easiest to hear. Above and below this frequency, it takes higher sound levels to achieve a given loudness. Low frequencies below 100 Hz are particularly hard to hear. It should also be noted that the curves become flatter with increased level.

Filter circuits were added to sound-level meters in an attempt to make them into loudness meters. Figure 4.17 shows the characteristics of filters developed in the 1930s and found in most sound-level meters today (ANSI S1.4). The three different filters were provided to account for the changes in the shape of the loudness curves with level. The A weighting was intended for levels below 55 dB, B for levels between 55 and 85 dB, and C above 85 dB. The practice that has evolved is to use only the A weighting. One reason is that it will generally rank the loudness of similar types of noises in the same order as a listening panel. The C weighting is sometimes used in

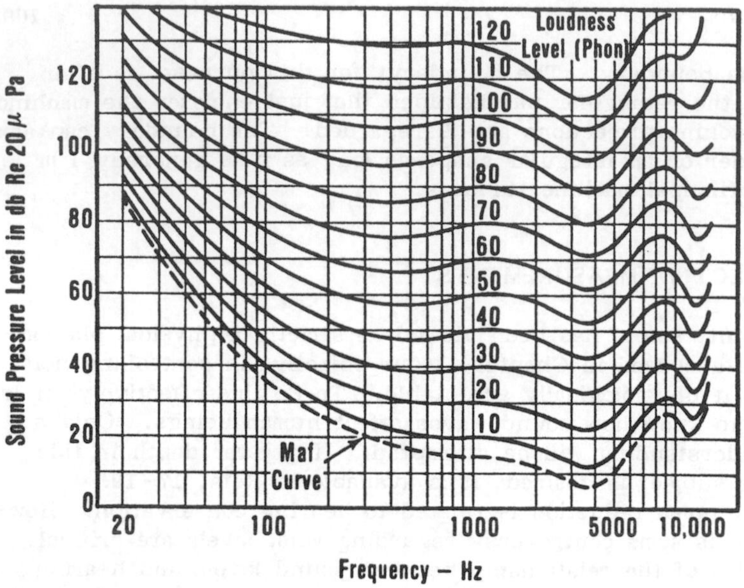

FIGURE 4.16 Loudness of pure tones. (Data from ISO R226.) This material is reproduced with permission from International Organization for Standardization Recommendation Normal Equal-Loudness Contours for Pure Tones and Normal Threshold of Hearing Under Free Field Listening Conditions, ISO R226-1961, © by the American National Standards Institute, New York.

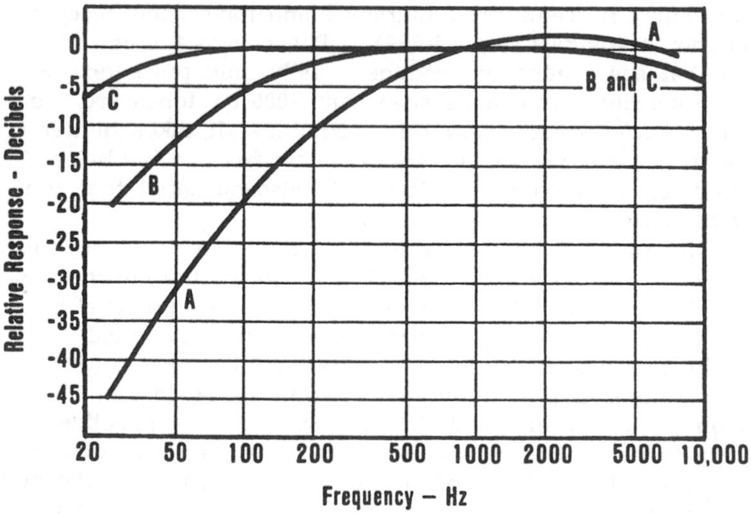

FIGURE 4.17 Sound-level meter filters. (Data from ANSI S1.4.)

conjunction with the A because the difference in their results indicates the nature of the frequency distribution of the noise being measured (20).

The designation dB(A) is used for measurements made with A weighting. This is pronounced "deebeeaye," with the last syllable rhyming with "hay." Although the form "dBA" would be easier on typists, its use is discouraged because it does not follow the units convention of the SI system.

Other weightings, such as N, D, E, and SI, have been developed in more recent times. These appear to be better measures of loudness or similar noise characteristics for some types of noise (21). For industrial noise, however, the A weighting scale has gained the widest acceptance and appears to provide a fair approximation of loudness from a single measurement.

A number of methods have been developed for computing the loudness of a noise from measurements of the noise made in narrow bands of frequencies. There are also methods for calculating perceived noise level, noise pollution level, speech-interference level, noise and number index, as well as a noise criterion (18). They are similar in nature and some are better in rating certain types of noises than others. They find very limited use in industrial and mobile machine noise control, however.

4.5.2 Loudness Calculations

In the 1960s I always converted noise measurements into loudness ratings. The advantage was that the scale was linear, a 40-sone noise sounding twice as loud as a 20-sone noise. The trouble with this rating is that it took about 2 days to convert 1 day's measurements. When Walsh-Healy used A weighting as the basis for workplace noise limits, it had the effect of consecrating this rating, and I was glad to quit calculating loudness.

A brief review of a loudness rating system is in order for the insights into the hearing process that it provides. This method was developed by S. S. Stevens of Harvard University (19). It went through a series of revisions, and one of these, referred to as the Mark VI version, was incorporated in both U. S. and international standards (ANSI S3.4 and ISO R532).

Figure 4.18 gives the data used to convert 1/3-octave band sound pressure levels into loudness indices in *sones*. It should be noted that it takes an increase of about 10 dB to double loudness.

The loudness of the total noise is found by adding the sones for the individual sound pressure levels in the following way:

$$\text{total loudness } S = I_m + 0.15(\Sigma I - I_m) \quad \text{sones}$$

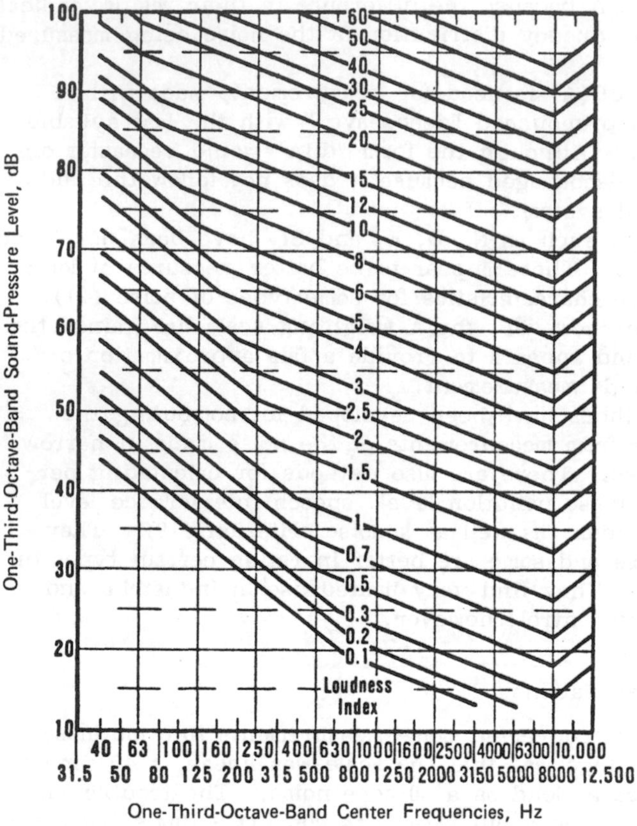

FIGURE 4.18 Level-to-loudness conversion chart. (Reprinted from ANSI S3.4, 1980, Computation of the Loudness of Noise by permission of the Standards Secretariat, Acoustical Society of America, New York.)

where

I_m = highest loudness index
Σ I = summation of loudness indices of all 1/3-octave band levels

From this we see that the loudest noise component is given much higher weight than all others. This is done to account for the fact that our hearing tends to focus on the loudest noise, Other noise components, even if they are nearly as loud, are "masked" by the loudest.

4.6 NOISE MEASUREMENT STANDARDS

American National Standards Institute
1430 Broadway, New York, NY 10018
(also available from Standards Secretariat,
Acoustical Society of America,
335 E. 45th St., New York, NY 10017)

ANSI S1.4, *Sound-Level Meters*

ANSI S1.11, *Specification for Octave, Half-Octave and Third-Octave Filter Sets*

ANSI S1.13, *Methods for the Measurement of Sound Pressure Levels*

ANSI S1.33, *Engineering Methods for the Determination of Sound Power Levels of Noise Sources in a Special Reverberation Test Room*

ANSI S1.34, *Engineering Methods for the Determination of Sound Power Levels of Noise Sources for Essentially Free-Field Conditions over a Reflecting Plane*

ANSI S3.4, *Computation of the Loudness of Noise*

International Organization for Standardization
(available from ANSI and ASA)

ISO R226, *Normal Equal-Loudness Contours for Pure Tones*

ISO R-532, *Method for Calculating Loudness Level*

ISO 3743, *Acoustics—Determination of Sound Power Levels of Noise Sources: Engineering Methods for Special Reverberation Test Rooms*

ISO 3744, *Acoutics—Determination of Sound Power Levels of Noise Sources: Engineering Methods for Free-Field Conditions over a Reflecting Plane*

ISO 4412, *Hydraulic Fluid Power—Test Code for the Determination of Airborne Noise Levels:*
Part 1. Pumps
Part 2. Motors

British Standards Institution
2 Park St., London W1A 2BS, England

BSI 5944, *Measurement of Airborne Noise from Hydraulic Fluid Power Systems and Components:*
Part 1. Method for Test for Pumps
Part 2. Method for Test for Motors

Part 3. Guide to the Application of Part 1 and Part 2
Part 4. Method of Determining Sound Power Levels from Valves Controlling Flow and Pressure
Part 5. Simplified Method of Determining Sound Power Levels from Pumps Using an Anechoic Chamber

National Fluid Power Association
3333 N. Mayfair Rd., Milwaukee, WI 53222

NFPA T3.9.12, *Method of Measuring Sound Generated by Hydraulic Fluid Power Pumps*

NFPA T3.9.14, *Method for Measuring Sound Generated by Fluid Power Motors*

National Machine Tool Builders Association
7801 Westpark Dr., McLean, VA 22101

NMTBA, *Noise Measurement Techniques*

REFERENCES

1. W. D. Stanley, *Network Analysis with Applications*, Reston Pub., Reston, Va., 1985.
2. T. F. Hueter and R. H. Bolt, *Sonics*, Wiley, New York, 1955, p. 61.
3. L. L. Beranek, *Acoutics*, McGraw-Hill, New York, 1954, p. 119; also republished in 1986 by the American Institute of Physics.
4. Ref. 2, p. 65.
5. Ref. 2, Part 7.
6. I. L. Ver and C. I. Holmer, "Interaction of Sound Waves with Solid Structures," in *Noise and Vibration Control*, L. L. Beranek, ed., McGraw-Hill, New York, 1971, p. 292.
7. F. Fahy, *Sound and Structural Vibration, Radiation, Transmission and Response*, Academic Press, Orlando, Florida, 1985.
8. G. Reethof and W. C. Ward, "A Theoretically Based Valve Noise Prediction Method for Compressible Fluids," *ASME Journal of Vibration, Acoustics, Stress, and Reliability in Design* 108: 329–338, July 1986.
9. A. C. Fagerlund and D. C. Chou, "Sound Transmission Through a Cylindrical Pipe Wall," *ASME Journal of Engineering for Industry* 103:355–360, Nov. 1981.

10. P. Hope, "Airborne Noise Radiated by Pipes and Other Components of Fluid Power Circuits Due to Internal Fluid Borne Noise," Report BH22, BHRA, The Fluid Engineering Centre, Cranfield, Bedford, England, Jan. 1979.

11. Ref. 6, p. 291.

12. Ref. 6, p. 293.

13. Ref. 6, p. 275.

14. R. Huntly, "A Bel Is Ten Decibels," *Sound and Vibration*, 4:22, Jan. 1970.

15. A. P. G. Peterson, *Handbook of Noise Measurement*, 9th ed., General Radio Co., Concord, Mass., 1980, p. 9.

16. R. A. Heron and I. Hansford, "The Development of a Method for Measuring the Sound Power Generated by Oil Hydraulic Relief Valves," paper C385/80, *Proceedings of the Quieter Oil Hydraulics Seminar*, Institution of Mechanical Engineers, Oct. 1980, pp. 77–85.

17. Ref. 15, "Hearing Damage from Noise Exposure," Chap. 3, pp. 17–24, and "Other Effects of Noise," Chap. 4, pp. 25–74.

18. L. L. Beranek, "Criteria for Noise and Vibration in Communities, Buildings, and Vehicles," in *Noise and Vibration Control*, L. L. Beranek, ed., McGraw-Hill, New York, 1971, Chap. 18, pp. 554–600.

19. S. S. Stevens, "Procedure for Predicting Loudness: Mark VI," *Journal of the Acoustical Society of America* 33:1577–1585, Nov. 1961.

20. J. H. Botsford, "Using Sound Levels to Gauge Human Response to Noise," *Sound and Vibration* 3(10):16–28, Oct. 1969.

21. Ref. 15, "Weighted Sound Levels," Sec. 4.21, pp. 66–69.

5

Pump Airborne Noise

Pump airborne noise reduction has received considerable attention. As a result, there are a number of reduced noise pumps on the market today. Also, fortunately, airborne noise is relatively easy to measure and the measurement procedure has been standardized (1). So most manufacturers supply standardized data for their products, making it relatively easy to compare noise levels for many of the pumps being sold today.

Airborne noise is only one form of pump noise. However, many of the steps that are effective in reducing it are also effective in reducing the other noise forms. Because of this the pump airborne noise rating tends to be an indicator of the other pump noises as well. This is particularly true when two similar pumps are being compared.

Reducing pump airborne noise is sometimes very difficult and is almost always expensive. It is particularly difficult if the pump has already been quieted to some degree. In such cases further quieting requires carefully reengineering many of the pumps design details and some cutting and trying. It also requires an extensive noise measurement program. Noise reduction modifications can adversely affect endurance, so additional cost is incurred in rerunning endurance qualification tests. Since port quieting changes sometimes increase volumetric efficiency slightly, retesting performance is also advisable. When the cost of tooling and other production changes are added, the bill for quieting a pump is generally very high. Although this chapter deals with quieting pumps, most of it is applicable to quieting hydraulic motors as well.

5.1 AIRBORNE NOISE SPECTRA

Figure 5.1 shows typical pump airborne noise spectra with peaks corresponding to the pumping frequency and many of its harmonics. The most notable feature is that some of the higher harmonics are stronger than the fundamental. In industrial and mobile pumps the highest harmonics are generally in the range 500 to 2000 Hz, as seen in spectra (a) and (b). Aircraft pumps sometimes peak at higher frequencies, as in spectrum (c). Another feature that must be noted is that sometimes the noise levels drop off rapidly at frequencies above the peak, as in spectrum (b).

Most of a pump's airborne noise energy is concentrated in a few of the strongest harmonics in its spectrum. Since these occur at the frequencies that are the easiest to hear, they have added importance because of the A weighting in sound measurements. It is not possible to reduce the total noise significantly unless these highest components are attenuated. For this reason pump noise reduction efforts are best concentrated on identifying the factors that can be altered to reduce the highest of these harmonics.

We know that in the Fourier series of the noise-producing forces and moments discussed in Chapter 2, the fundamental component always predominates, and strengths tend to decrease with harmonic number. The differences between force spectra and those in Figure 5.1 provide clues to the other mechanisms involved in producing pump noise. Three mechanisms are active. Although it is not known exactly how each contributes in a specific case, there is ample evidence that all three are at work.

In earlier chapters we saw how sound waves are generated by simple vibrating sound sources. Here, we focus on radiation from pump housing vibrations. Pressure within the passages in the valve plate of a piston pump, for example, causes a localized area to pulse like a very stiff diaphragm. Acoustically, this is similar to the piston in a baffle discussed in Section 4.1.3. Similarly, axial force variations in vane and gear pumps cause their length to fluctuate accordingly. The piston forces acting between the valve and swash plates do the same thing to piston pumps. These pump elongations produce sound in about the same way as a pulsing sphere.

5.1.1 Velocity Factor

Based on the simple vibrating system discussed in Chapter 3, we generalize that the motion of these housing surfaces, in response to a sinusoidal force, $P = p \sin \omega t$, is

$$x = \frac{(DF)p \sin \omega t}{K} \quad \text{in.}$$

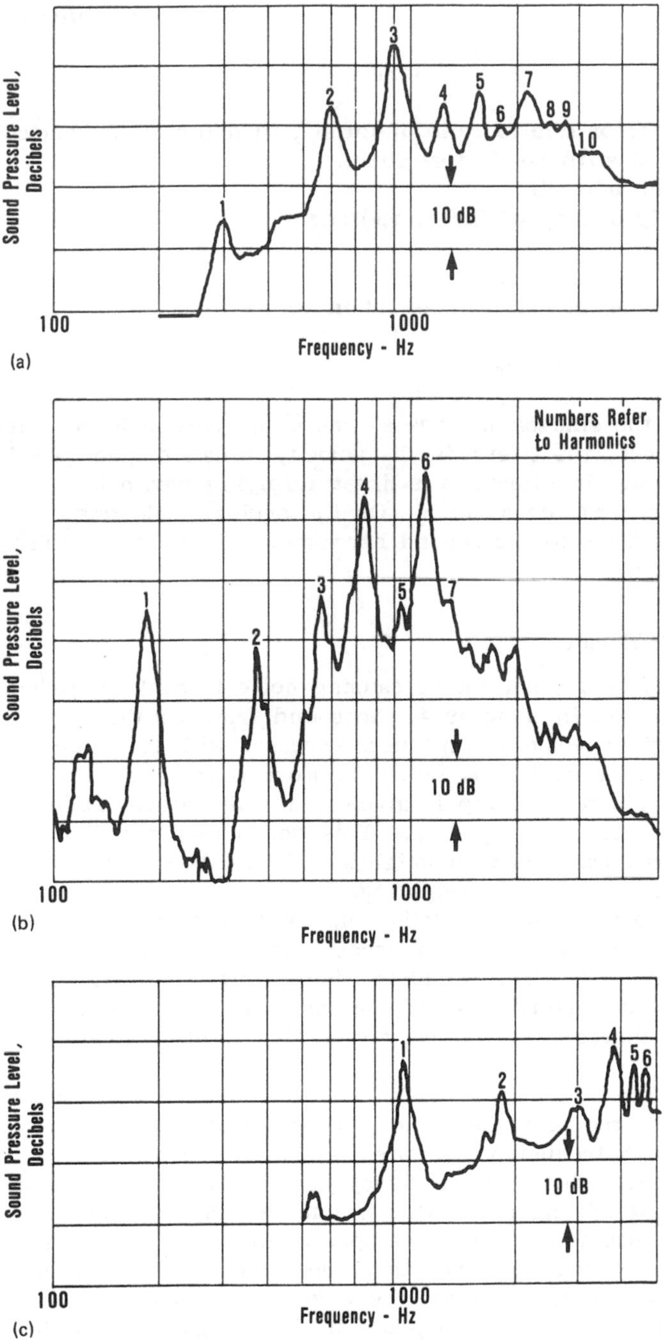

FIGURE 5.1 Typical pump airborne noise spectra: (a) 10-vane pump, 1800 rev/min; (b) nine-piston pump, 1800 rev/min; (c) seven-piston pump, 8000 rev/min.

where

DF = dynamic factor due to ratio of forcing to natural frequencies
K = generalized stiffness factor, lb/in.
p = force amplitude, lb
ω = angular frequency of force, rad/sec

Sound radiation is proportional to surface velocity. Velocity is found by differentiating the displacement equation above

$$\dot{x} = \omega(DF)p \cos \omega t \quad \text{in./sec}$$

If we think of the sinusoidal force as one component in a harmonic series of such forces, the ω term in the velocity equation gives weight to higher harmonics. In effect, it is like multiplying harmonic strengths by their harmonic number. On the decibel scale this amounts to adding 3 dB to the second harmonic, 4.8 dB to the third, and 6 dB to the fourth.

5.1.2 Radiation Efficiency

Radiation efficiency is another factor causing noise to peak at higher harmonics. As we saw in Chapter 4, the sound from a radiator vibrating with a given velocity amplitude increases with frequency until a maximum efficiency is reached. For most of the radiation models for which this relationship is given in the literature, radiation increases as the square of frequency. This means that radiation efficiency provides harmonics with weighting, in decibels, which is twice that provided by velocity weighting.

One exception to this square relationship for radiation efficiency is the piston operating in free space, whose radiation characteristics are shown in Figure 4.8. The radiation efficiency of this model is extremely poor at low frequencies and increases more rapidly with frequency. This characteristic gives very high weighting to the higher harmonics.

In the case of the piston in a baffle, shown in Figure 4.6, radiation increases with frequency until the wavelength of the sound, which decreases with frequency, becomes less than one-half the piston circumference. Radiation efficiency does not increase beyond this point, so the weighting above this frequency stays constant. Examination of a number of pump noise spectra indicates that large pumps level off at lower frequencies than small ones. The leveling off appears to occur, roughly, when the wavelength equals half the pump circumference, measured in a plane at right angles to the shaft axis. This should not be taken as suggesting that the pump is acoustically similar to the piston model, but only as a very rough rule of thumb.

5.1.3 Dynamic Factor

The combination of the velocity and radiation weighting added to the decreasing strength of the forces appear to account for some of the higher harmonics being strongest in many pump noise spectra. However, it does not account for the rapid drop-off above the peak, like the one in Figure 5.1(b). This rapid drop-off is the strongest evidence that the dynamic factor of the vibration equation, given in Section 5.1.1, is also a factor in shaping pump airborne noise spectra.

This drop-off results from having a harmonic excite resonance in the pump structure. This resonance does not have to be in the housing where the sound is radiated. It can be anywhere along the path from where the forces or moments are generated to the surface where radiation occurs.

The technique used to verify the existence of such resonances is to make sound measurements at a series of speeds spaced about 100 rev/min apart. Large changes in a harmonics strength indicates resonance.

Resonance does not appear to be a frequent factor in pump noise. It is not known if this is because it seldom occurs at frequencies where strong excitation exists or because it is usually highly damped and is not detected because it provides little amplification.

5.2 EFFECT OF PORT TIMING

The greatest reductions in pump airborne noise were made with port timing improvements. These modify the noise-generating forces and moments so that their most important harmonics are reduced, as discussed in Section 3.3. This is discussed in greater detail later in this section, but first it is best to examine how such changes are effected.

5.2.1 Operational Effects

The time profiles of pump forces and moments depend on pump speed and discharge pressure as much as they do on port timing. This was learned from the first pump that the author was involved in quieting. A port timing was found by cut and try that made the pump remarkably quiet at rated speed and pressure. When the pump was introduced at a hydraulic trade show, cooling water was not available, so it was operated at zero pressure. At this condition it was heard throughout the exhibition hall and had to be withdrawn.

Since most pumps operate over a wide range of conditions, porting has to be a compromise that yields acceptable noise levels over this range. Lower levels are possible at a single set of conditions, but this leads to higher levels at other conditions. Maximum quieting through port timing is possible only when a pump operates within a narrow range of conditions. Generally, because of cost, it is not practical to develop quiet timing for single applications, so most pumps are developed for operation over their catalog-specified ranges.

Certain portions of the operating range are favored by making pumps quieter at these conditions. Some companies favor rated pressure because it results in more favorable advertising copy. Others feel that customer satisfaction is greater if noise increases slightly as pressure increases. Some try to attain nearly the same level at all conditions. All strategies must avoid undue levels at some condition within their operating range.

5.2.2 Computer Modeling

The most useful tool for pump quieting is a computer model that computes pumping chamber pressure-time profiles. A program of this type makes it fast and easy to examine the effects of port timing changes at a number of operating conditions. The cost of developing the program is quickly recovered by the reduction in the testing required to achieve noise reduction goals.

A model considers fluid flow into and out of the pumping chamber as well as changes in the chamber volume. Generally, only flows through orifices need to be considered. Leakage through clearances can be ignored in piston and vane pumps but may have to be considered in gear pumps. Step pressure changes are computed by assuming that flows remain constant over a short interval while the chamber volume changes an amount commensurate with the time interval. Flows in the next interval consider the newly computed pressure and changes in the chamber orifice areas. Very small intervals must be used or the calculated pressures will oscillate widely over and under the real pressure. In some cases it is necessary to use 0.01° increments to avoid instability.

Inline piston pumps probably are the easiest to model because cylinder volume is a simple sine function. Ernst (2), Hadekel (3), Thoma (4), and Wilson (5) are good references for modeling other pump types. For units such as balanced vane pumps, in which all pumping chamber cross sections change with shaft rotation, the model must calculate a number of equally spaced cross sections of the pumping chamber and then integrate them to determine the chamber volume for each calculation interval. This is tedious, even with a computer, so care must be taken to limit calculations strictly to the pressure transition portions of the pumping cycle.

Computer models of inline piston and balanced vane pumps have been verified with small pressure transducers that measured actual pumping pressure profiles. Although they are very difficult to make, such measurements are highly recommended for the detailed insights that they provide. They are not necessary for effective modeling, however.

Bulk modulus, which is a key parameter in pump pressure compute models, is discussed in detail in Section 6.1.2. For our purposes here we note that an effective bulk modulus of hydraulic oil of 180, 000 psi works well in models for commercial pumps operating at pressures up to 3000 psi. Orifice coefficients for metering grooves and other small flow paths of 0.65 also give good results.

Some cut and try testing is in order no matter how well refined the model and how many different sets of operating conditions are evaluated. When computations indicate a timing is promising, it is best to make the first test porting so that the computed timing is achieved by removing additional metal. After the first porting is tested, it is modified to the desired configuration and retested. Additional modification and retesting are then advisable. This procedure will establish if the selected porting is optimum, or if not, the direction to the optimum.

Fortunately, noise measurements do not require extensive pump operation, so experimental portings are usually tried in unhardened port plates to save time and money. This is good because if a porting causes cavitation, this is indicated by plate erosion in the noise tests. Changes are not limited to those requiring metal removal, as brazing can be used to fill in grooves and even portions of the ports themselves.

5.2.3 Effect of Port Detail

Earlier in this chapter it was seen that pump airborne noise is due primarily to forcing harmonics in the range 500 to 2000 Hz. In Section 3.3 we saw from Fourier analyses of trapezoidal waves, which roughly approximate some forcing functions, that increasing transition time decreases the strengths of higher harmonics. The objective of changing port timing is therefore, to extend pumping chamber pressure transition times and thereby reduce the strengths of forcing harmonics in the range 500 to 2000 Hz. Actually, this is a simplification of the strategy, but is adequate until we have the opportunity to refine it, later in this chapter.

One of the quietest port timings is achieved by delaying the communication between the pumping chamber and the pump ports until cyclic variations in the chamber volume adjust the chamber pressure to that of the next port. To illustrate this process we use a hollow-piston axial piston pump for an example.

Many industrial and mobile machinery piston pumps have clearance volumes of about 1.7 times their swept volume. Most of this volume is within the hollow pistons. In Figure 2.2 the cylinder volume at bottom dead center where the cylinder commutes from the inlet to the discharge port is

$$V = 2.7(2Ar \tan \beta) \quad \text{in./sec}$$

where

A = cylinder cross-sectional area, in.^2
r = cylinder base circle radius, in.
β = swash plate angle, deg

As the cylinder rotates past bottom dead center, the change in this volume is

$$\Delta V = Ar \tan \beta(1 - \cos \theta) \quad \text{in.}^3$$

where θ is the cylinder rotation, measured from bottom dead center, in degrees. When communication with the discharge port is delayed, the cylinder pressure is

$$P = \frac{B\,\Delta V}{V} = \frac{B(1 - \cos \theta)}{5.4} \quad \text{psi}$$

where B is the oil bulk modulus in psi. The delay in opening to the discharge port to equalize the cylinder pressure to a given discharge port pressure is

$$\theta = \cos^{-1}\left(1 - \frac{5.4P}{B}\right) \quad \text{deg}$$

If we use 180,000 psi for bulk modulus, the delay needed for a discharge pressure of 2000 psi is 19.9°.

Similarly, to decompress the cylinder the inlet port opening is delayed

$$\theta = \cos^{-1}\left(1 - \frac{3.4P}{B}\right) \quad \text{deg}$$

Because there is less fluid in the cylinder at this point, a delay of only 15.8° is required to decompress from the same pressure. This porting is shown in Figure 5.2.

This is a neat solution for pumps that always operate at the design pressure of 2000 psi. Unfortunately, most pumps do not. A pump with such porting will be loud at 1000 and 3000 psi and very loud at 0 psi. Simple delayed portings are very intolerant of deviations from their design pressure.

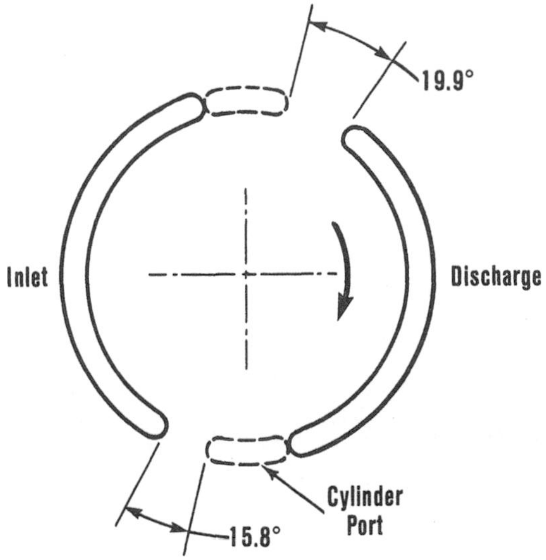

FIGURE 5.2 Delayed porting for 2000 psi.

Metering grooves such as the one shown in Figure 5.3 are generally used in combination with delayed porting. Less delay is needed when they are used because they provide metered flow to assist in equalizing pumping chamber pressure. Since this flow is a function of the pressure differential, it makes the porting more forgiving of discharge pressure variation. When the combination of metering groove flow and chamber volume change provides excess pressure, the flow reverses direction to reduce the excess.

Pressure rise through metering groove flow is time dependent. So, unlike simple delayed porting, metering groove portings are sensitive to pump speed as well as pressure. Since they equalize pressures with flow added to volume changes, pressure transitions at the design conditions are always faster than with delayed porting alone. For this reason, metering grooves do not provide the maximum quieting achieved with delayed porting. Even with these disadvantages, however, they are widely used because they provide more quieting when operating at conditions other than those for which they were optimized.

The flow into or out of pumping chambers required for pressure transitions is very minute. Many metering grooves seen in pumps today provide this flow within a few degrees of shaft rotation. They therefore provide only a small percent of their noise reduction potential. The most effective grooves are long and shallow.

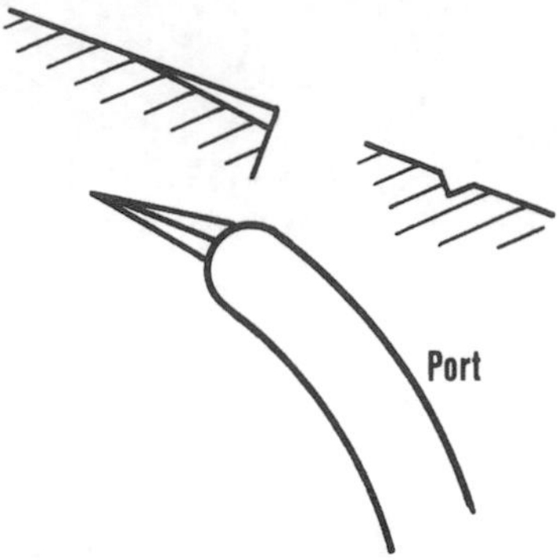

FIGURE 5.3 Typical metering groove.

5.2.4 Cavitation Effects

Cavitation is the archenemy of pump quieting. When cavities or bubbles are present in a pumping chamber as compression begins, no pressure rise occurs until flow and volume changes fill them. This delays compression and it is not complete when the pumping chamber opens to the discharge port. This makes the pump noisy. In extreme cases where cavities still exist at port opening, fluid from the port rushes in and causes extremely rapid bubble collapse. This produces a sound that is like gravel being crushed in the pump.

Cavity collapse itself generates considerable noise energy. Research studies suggest that some of this energy is due to micro diesel combustion that occurs because the cavities are filled with a mixture of hydraulic fluid vapor and air. Even without the diesel effect, cavity collapse produces enough energy to cause serious metal erosion in pumping chamber and porting surfaces.

Cavitation is sometimes due to air entrained in the fluid supplied to the pump. Eliminating this cause is outside of pump design and is discussed in Section 12.3. Pumps are capable of generating their own cavitation, and this is our concern here.

Cavitation occurs when fluid pressure drops below the saturation pressure of the air dissolved in the fluid. This is promoted by inlet pressure losses, but high fluid velocities are the biggest cause. Vortices generated when fluid flows around obstructions in inlet passages have very low pressures at their centers and are a common source of cavitation. Flow is accelerated to high velocities in orifices and we find that even metering grooves are cavitation sources.

High-pressure pumps are particularly susceptible to cavitation because with their high-pressure differentials they generate very high metering groove velocities. As a matter of fact, cavitation severely limits the amount of porting noise reduction that can be designed into high-pressure pump porting.

It is felt that quiet high-pressure pumps are not possible until some method for suppressing cavitation is found. Although no adequate control appears to be on hand, several things are found to help and raise hopes for the future. For example, the presence of solid surfaces retards cavitation. Multiple, shallow, and parallel metering grooves cavitate less than single grooves of the same total flow capacity. Similarly, less cavitation develops when pressure drops occur over longer distances. In valves, cavitation is suppressed by cascading orifices in series so that the drops across each is lower. No way to apply this to pumps has yet been found, but it could be the key to the future. Supercharging at the pump inlet is another cavitation control that is worth consideration from a noise standpoint.

Some cavitation occurrences appear to contradict the foregoing discussion. For example, a short and relatively deep metering groove was found to cause less cavitation erosion than a longer shallow groove. This was taken to indicate that the deeper groove resisted cavitation. However, this groove is cut at a steeper angle to the valving surfaces, so its jet is directed farther from the wall of the pumping chamber, where the erosion occurred. It is believed that the erosion difference is the result of fewer cavities collapsing at this surface. Neither groove made the pump quiet at the conditions that caused erosion.

Another case is a common one. Under some operating conditions, a metering groove used to control decompression at the start of the inlet port caused erosion at the discharge port metering groove. Erosion disappeared when the decompression flow was directed into the pump case instead of the inlet port. This was not a practical solution, but it did demonstrate that the decompression metering groove jet was causing cavitation in the inlet port. The bubbles were then carried to the start of the discharge port, where increasing pressure caused them to collapse and erode the surface. This condition was alleviated by providing more delayed port decompression.

5.2.5 Pressure Profile Analyses

When a computer program for calculating pumping chamber pressure cycles is used, calculations must be made for a number of different sets of operating parameters for each porting considered. The problem then becomes one of deciding if the improvement seen at one condition outweights poorer results at another condition. This would be relatively easy to decide if transition duration was the only parameter that affected noise, as suggested earlier in this chapter.

Instead, the problem expands into wondering if "smoothness" of transitions and the presence of pressure overshoot must also be considered.

Such questions were studied by making Fourier analyses of a series of idealized pressure profiles. The results provide guidance for making judgments by inspection. Although Fourier analysis is not particularly difficult with a computer, it probably is not warranted for each porting study.

Analyses were made for profiles that might occur in a piston pump operating at 3000 psi, but to simplify the arithmetic, the profiles were normalized to a pressure of 1 psi. It was assumed that the clearance volume was 1.7 times the swept volume, so the cylinder volume at decompression was 63% of that during compression. The time required for pressure transitions are therefore inversely proportional to these volumes.

Idealized rather than calculated profiles were used so that the effect of individual profile characteristics could be evaluated separately. These simulations are sufficiently realistic to make the results useful in actual cases, however.

A nine-cylinder pump was assumed and the profiles were used to calculate the force and moment functions discussed in Chapter 2. Fourier analyses were then made of these force and moment functions. Forty intervals were used in the analyses to ensure that the effect of even fine detail would be evaluated. The strengths of only the first 10 harmonics were determined since higher harmonics are seldom significant. The piston force component strengths were normalized using the strength of the square-wave fundamental as a base. Since the fundamental of the square-wave yoke moment and torque are zero, their second harmonics were used for normalizing. Harmonic strengths are given in decibels, $10 \log(S/S_B)$, to make them comparable to sound pressure levels.

Pure delayed porting provides the maximum transition time that can be provided with conventional porting, so a profile for this porting type was analyzed first to provide a bogey for interpreting subsequent studies. This profile and the harmonics of its forcing functions are shown in Figure 5.4.

The effect of pressure rise rate was studied first, by analyzing profiles with compression times varying from 0 to 20°, in 4° increments. Decompression times are proportionally shorter.

These trapezoidal profiles and their force function spectra are shown in Figure 5.5. In some cases functions have alternating high and low harmonics, which cause their plots to crisscross those of others. To reduce the confusion in such cases, only their high harmonic levels were plotted since these are considered to be of the greatest importance.

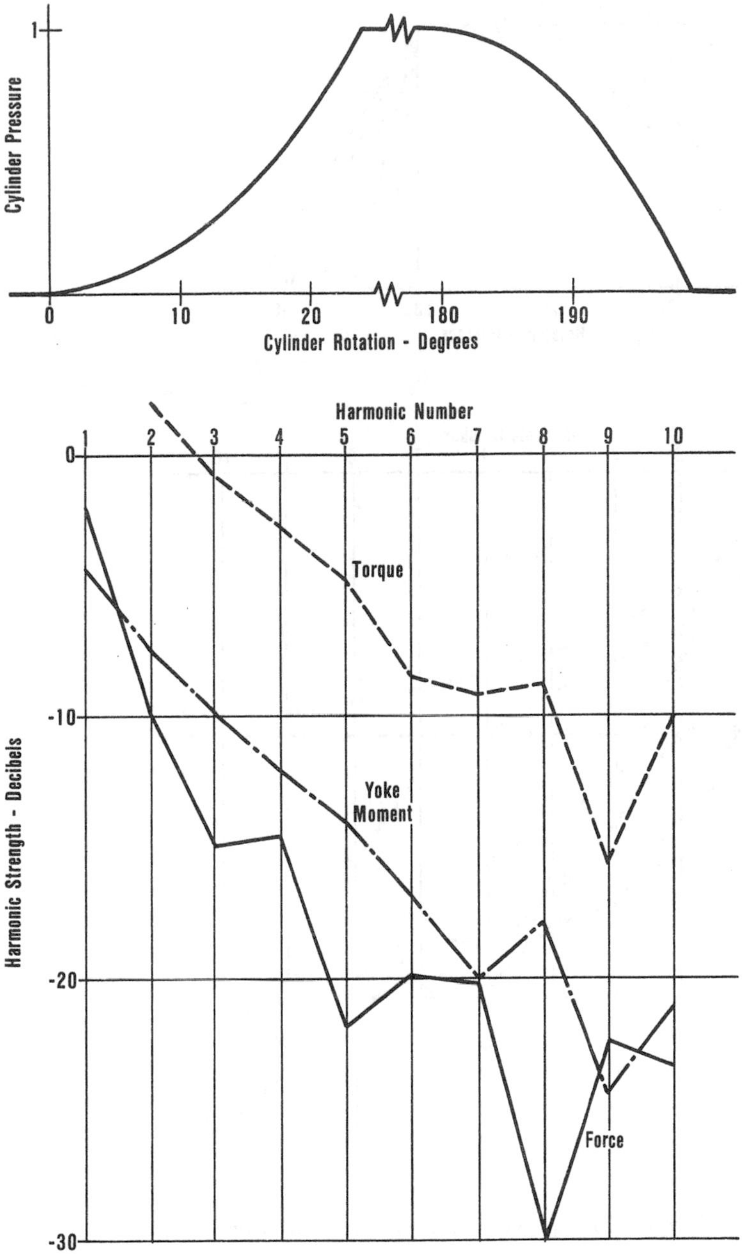

FIGURE 5.4 Pressure profile and function harmonics for delayed porting.

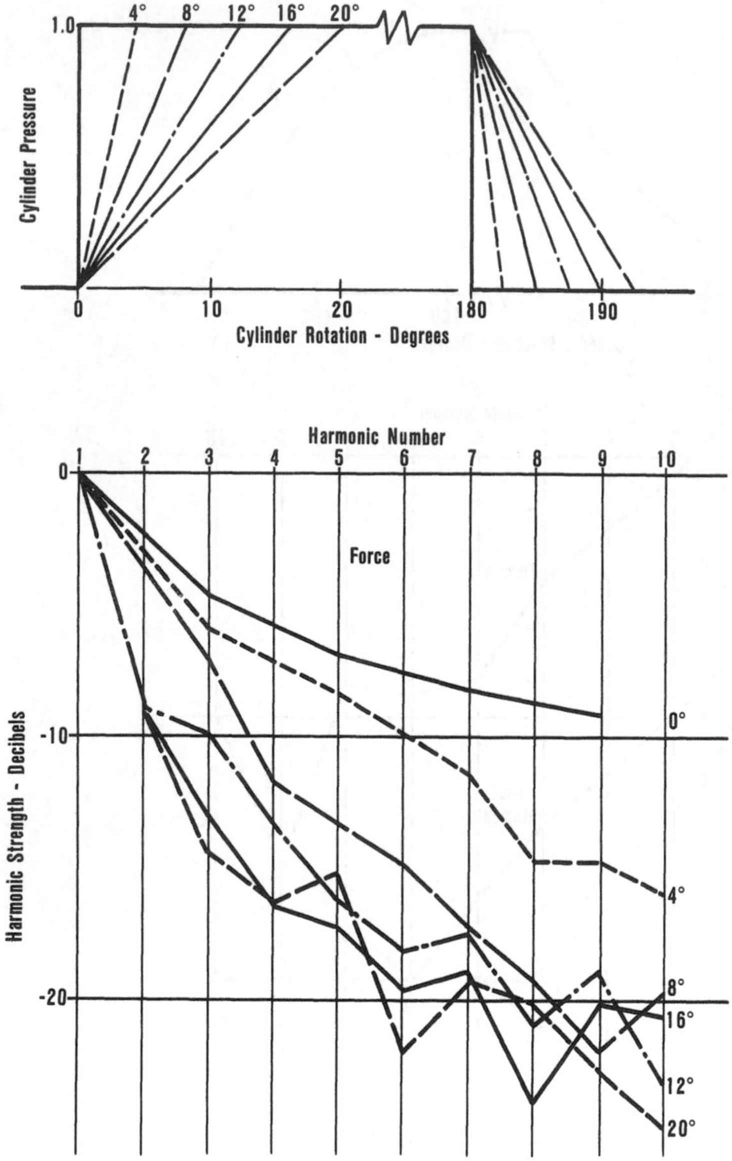

FIGURE 5.5 Trapezoidal pressure profiles and force function harmonics.

It can be seen from Figure 5.5 that increasing transition time is very effective in reducing the piston force harmonics. The rate of improvement decreases as transition time increases, so that the gains after 8° are small. Above this, the decrease in some harmonics may be offset by increases in the others. Comparison with Figure 5.4 shows that delayed porting is still noticeably better than even the 20° trapezoid.

Figure 5.6 shows the yoke moment harmonics for the trapezoidal pressure profiles. The effect of increasing rise time is similar to that for the force function. Benefits extend to higher rise times, however. It should be noted that when the rise time exceeds 12°, the fundamental frequency components become significant. Also, comparison with Figure 5.4 shows that the 16 and 20° profiles are better than delayed porting.

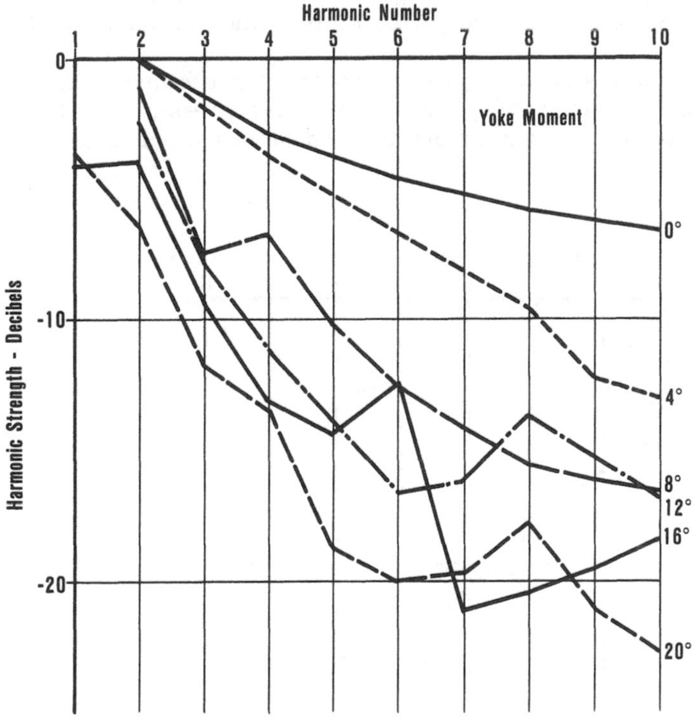

FIGURE 5.6 Yoke moment harmonics resulting from trapezoidal pressure profiles.

Torque harmonics of some of the trapezoidal pressure profiles are shown in Figure 5.7. These curves overlap each other so much that a plot of all the data becomes meaningless. From this it is concluded that increasing transition times has little effect on this function.

The trapezoidal profiles are made up of a series of straight lines. Measured and calculated profiles have rounded corners, so the second study evaluated the effect of corner curvature. Parabolas were used to blend between sides of the 12° rise-time trapezoidal profile of the first study. Three curvatures, as shown in Figure 5.8, were used. Their force harmonics and those of the 12° trapezoidal profile are shown in Figure 5.9.

It shows that smoothness is as important as slope. The same conclusion is drawn from inspection of yoke moment harmonics, which are not shown. Since the curves also extend the transition time, it may be useful to think that the benefits are due to this.

Inspection of the torque harmonics for these profiles shows that although the sharpest of the blending curves was beneficial, the larger curvatures were not much different from the profile without blending.

When porting is designed to provide adequate precompression at a given pressure, it generally provides excessive pressure for lower discharge pressures. This is referred to as pressure overshoot.

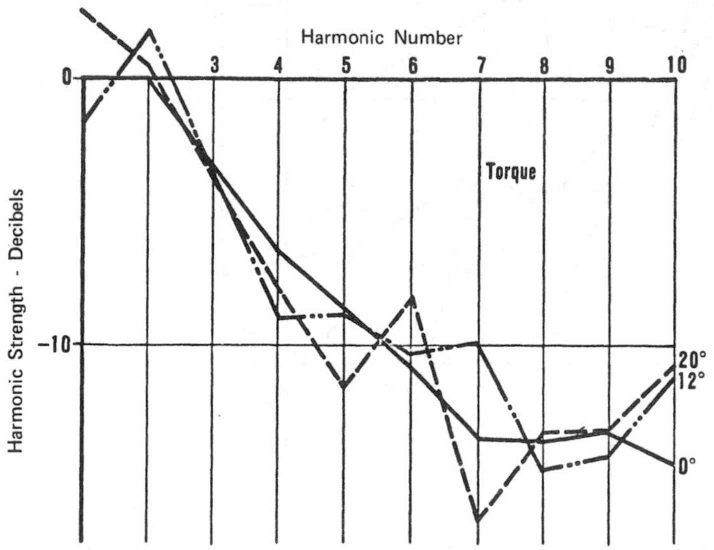

FIGURE 5.7 Trapezoidal pressure profile torque.

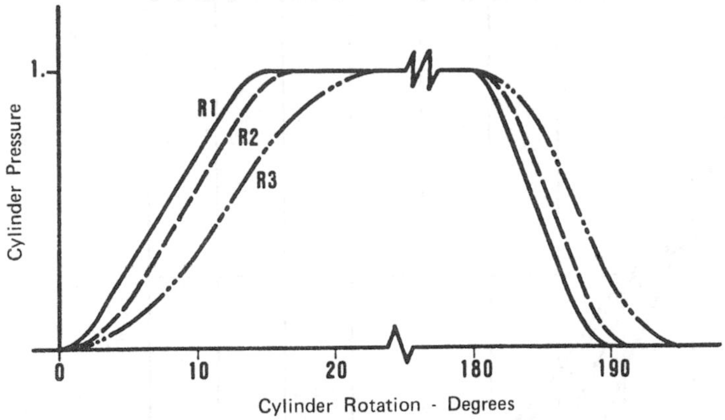

FIGURE 5.8 Smooth pressure profiles.

When a large overshoot occurs with simple delayed porting, the pressure drops suddenly as the port opens and high noise levels occur. This is well accepted. However, when metering grooves are used, these sometimes smooth out the overshoot so that the noise increase does not seem so certain.

A profile of this type, with 25% excess pressure, was analyzed. This curve, which was otherwise like the profile with the sharpest curvatures in the second study, is shown in Figure 5.10. Its effect on the force harmonics is shown in Figure 5.11. Instead of increasing noise, it is seen that the fourth and fifth harmonics are greatly reduced. Very little change occurred in the yoke moments. Torque was increased significantly, however, as seen in Figure 5.12. It is difficult to generalize on the effects of overshoots from these data. From experience they do not seem to have much effect when they are not excessive and they are well rounded.

5.2.6 Synthesizing Noise Spectra

It is useful to examine how force harmonics act on a pump structure to produce noise. To do this we add the response weightings, discussed in Section 5.1, to force harmonics to see if the results look like measured airborne spectra. This is a speculative exercise and is only academic. It is not a proof of the concepts being discussed, nor is it proposed as a noise reduction technique.

Figure 5.13 shows a noise spectrum created in this way. The force harmonic series for the overshoot pressure profile was used. A radiation critical frequency of 750 Hz, a natural frequency of 1100 Hz,

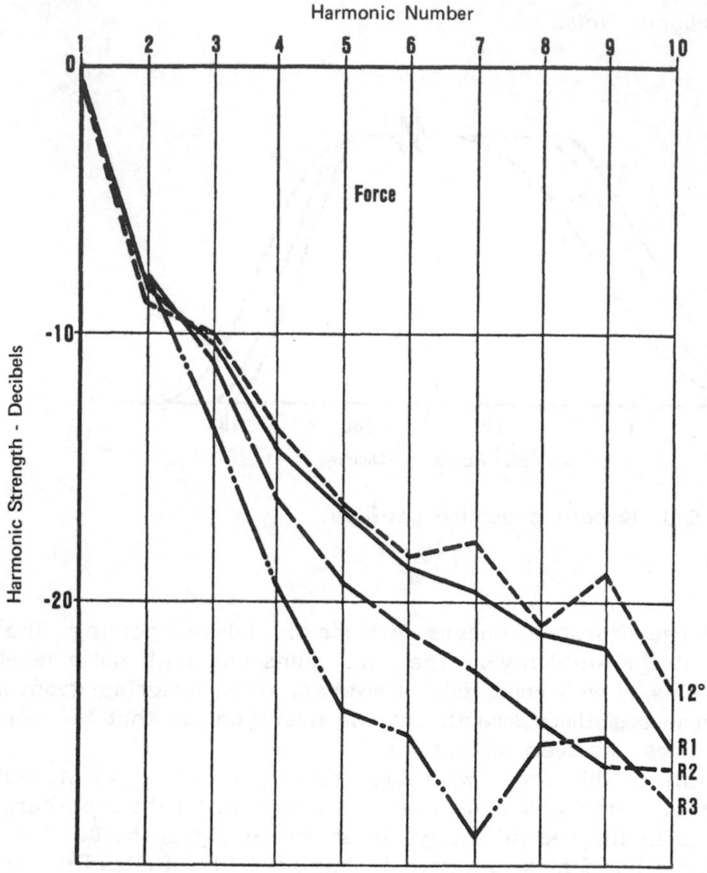

FIGURE 5.9 Effect of pressure profile smoothing.

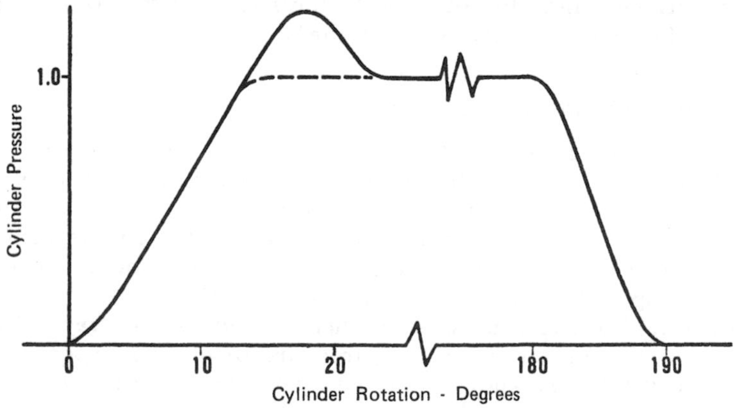

FIGURE 5.10 Overshoot pressure profile.

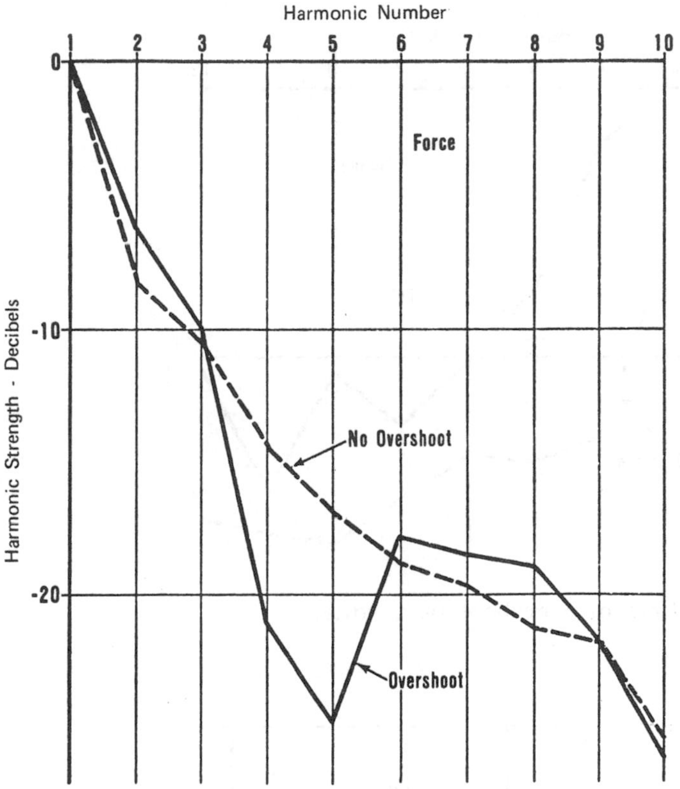

FIGURE 5.11 Effect of overshoot on force.

and a damping ratio of 0.2 were assumed. The selection was made by cut and try in an attempt to duplicate the spectrum in Figure 5.1(b) as closely as possible.

Figure 5.14 shows how the three types of weighting add to the forcing function harmonics to achieve the spectrum of Figure 5.13. It can be seen that all three factors make significant contributions to the final result. The radiation efficiency factor has the greatest effect. Whenever a spectrum tends to be relatively flat, resonance will be the smallest contributor.

This synthesizing process is also useful for visualizing the effect of pump speed. The small radiused pressure profile shown in Figure 5.8 was used for this study. Pressure profiles change with speed, especially if most of the pressure changes are due to metering groove

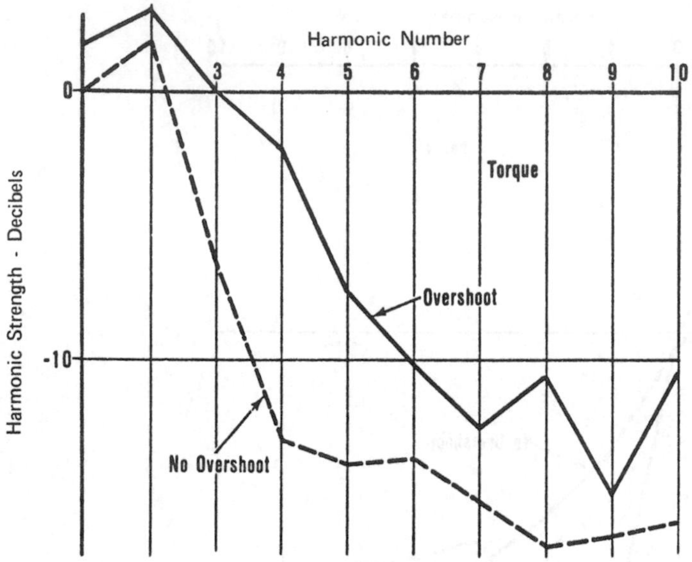

FIGURE 5.12 Effect of overshoot on torque.

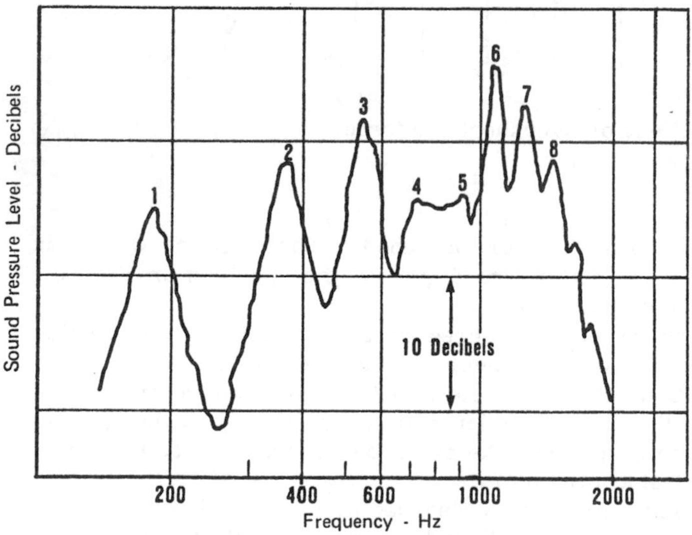

FIGURE 5.13 Synthesized noise spectrum.

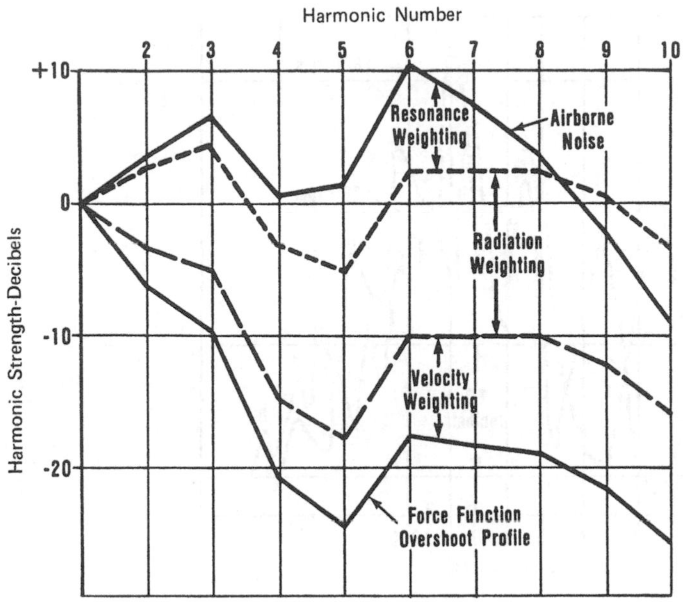

FIGURE 5.14 Synthesized spectrum weightings.

flows. However, for this study it was assumed that the profile remains constant, so that the main speed effect was more apparent.

A radiation critical frequency of 750 Hz, a natural frequency of 900 Hz, and a damping ratio of 0.2 were assumed. Spectra for 1200- and 1800-rev/min pump speeds were calculated and are compared in Figure 5.15. The strength of the 1200-rev/min fundamental component was used as the base for the decibel scale. At the higher speed harmonics occur at higher frequencies, and therefore those below the natural frequency are enhanced or weighted more. This comparison illustrates the major reason that speed has a major effect on pump airborne noise.

5.3 EFFECT OF PUMP CONFIGURATION

Some pump types have reputations for being inherently quiet. The screw pump is generally accorded this distinction. It is felt that its structure, alone, is not responsible for its observed low noise levels.

This opinion stems from a comparison made of a fixed-displacement piston pump and a screw pump of comparable displacement. It was

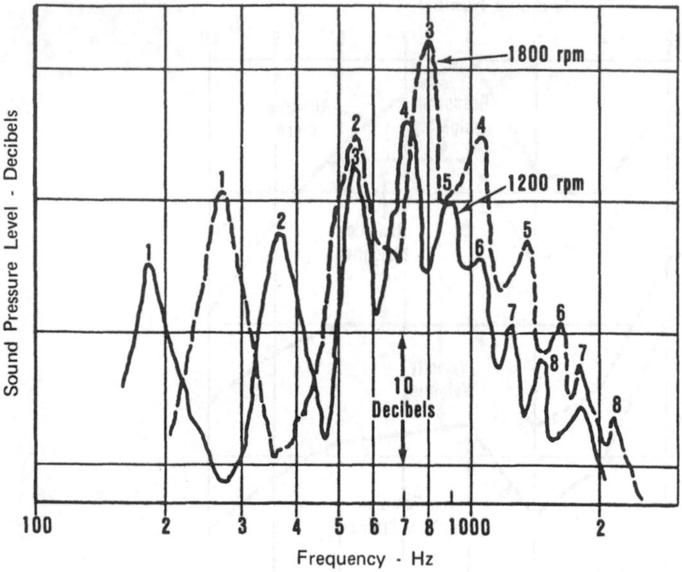

FIGURE 5.15 Effect of speed on airborne noise.

an experiment suggested by P. J. Wilson of Vickers Inc., U. K. The objective was to learn about the noise-generating process, not to determine which pump was superior. Unfortunately, the piston pump did not operate satisfactorily at 3600 rev/min and the screw pump's efficiency was unacceptably low below this speed. The comparison therefore had to be made at different speeds. It is felt that if the conditions making this necessary could have been eliminated, the comparison would still have a similar result.

Compression in a screw pump is accomplished by leakage into the pumping chamber. It is therefore very gradual. To try to duplicate this in the piston pump, shallow grooves were cut into the port plate, connecting the ports as shown in Figure 5.16. The ports were shortened to increase the groove lengths. Flow through the grooves set up linear pressure gradients which were sensed by the pumping chambers as they traveled between ports. Cylinder pressure change was therefore very gradual. The pump was as quiet as the screw pump and, coincidentally, had about the same volumetric efficiency.

From this it is concluded that differences in pump noise are due primarily to design details that are controlled by the designer. In most cases pumps are quiet because they are designed for low noise, not because they are inherently quiet.

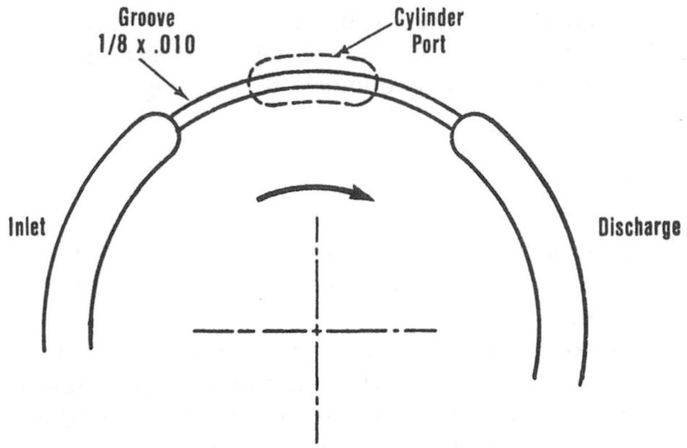

FIGURE 5.16 Pressure gradient porting.

Some pumps may be easier to quiet, however. In piston pumps the piston moves inextricably and so limits the duration of pressure transients. In balanced vane pumps, cam profiles can be made with sections where the pumping chamber remains constant, so that the time for pressure transitions is increased. However, these pumps have the disadvantage of having to complete their entire cycle in only one-half a revolution. Some gear pumps have almost this much time for their compression but must decompress very rapidly. We pointed out these differences to focus attention on the factors mentioned. With good engineering and perhaps some invention, such factors can be controlled to make a pump quieter than those reputed to be "inherently quiet."

There is one noise difference related to pump type that it may not be possible to eliminate. All variable-displacement pumps are louder than comparable fixed-displacement pumps. The reason for it is not understood, but intuitively it seems reasonable to believe that the difference is immutable.

5.4 IMPACT NOISE

The impact that concerns us here is cyclic, usually occurring at a specific point in each pumping cycle. Pump parts experiencing impact shock are set into free vibration like that discussed in Section 3.1.2. The resulting noise is primarily at pumping frequency and perhaps a few of its harmonics. The noise also has components at the ringing or natural frequencies of the impacting parts.

Serious cases of impact result in mechanical damage that pinpoints their cause. These are usually corrected during development to secure adequate pump life. There are boderline cases, however, that may escape detection but are still severe enough to add significantly to the pressure-generated noise.

5.4.1 Yoke Rattle

The beneficial effect on noise of increasing pressure transition times in piston pumps was discussed earlier. Increasing this time can have a negative noise effect as well.

Figure 5.17 shows the yoke moments resulting from some of the profiles analyzed in Section 5.2.5. It shows that increasing transition time in trapezoids reduces the positive yoke moment peaks considerably. Rounding the profile corners causes an additional change; it increases the negative peaks, which means that the yoke has a greater tendency to move off stroke. This is typical of most quiet timings.

Pump controls designed to provide adequate force for keeping the yoke on stroke with noisy timings may not provide enough with quiet timings. Enough force has to be applied to clamp the yoke during the peak negative moment or it will rattle against its stop. This produces a strong buzzing noise that may be louder than the pump with noisy timing. Usually, a higher return spring or bias piston load is required. In some cases this requires major design modifications.

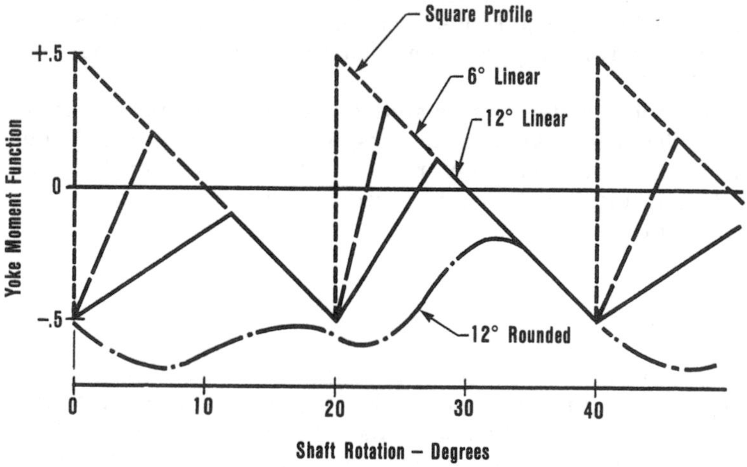

FIGURE 5.17 Effect of quiet timing on yoke moments.

Yoke rattle can also be the result of resonance. Compression of the fluid in the stroking cylinder is often the major spring element of the resonating system. Unlike most springs, its stiffness changes with its extension so that the resonance will generally occur at only one yoke position. The inertia of such systems includes the piston masses as well as the yoke moment of inertia. In the case of mobile machine pumps that are controlled through mechanical linkage, the inertia of the linkage also contributes.

The cure for this type of problem lies in "tuning" the vibration system. This is made difficult by the requirement of avoiding strong harmonics.

5.4.2 Piston Pounding

Cylinder pressure clamps the pistons, their shoes, and the swash plate together until the cylinder depressurizes. This occurs just when the piston is being accelerated outward to retract it for the suction stroke. The acceleration force comes from a retraction ring that presses against the back side of the shoes. The shoes, in turn, pull on the pistons. In addition to the acceleration force this system also supplies the force to overcome viscous friction drag on the piston, which is very high in high-speed pumps. Later during the suction stroke the acceleration force reverses. At this time or later, when pressurization begins, the piston, shoe, and swash plate are again clamped together.

If there is appreciable clearance between the parts, impact occurs each time the load reverses. This is usually a degenerative condition because the pounding increases the clearances and this further increases the impact forces. In extreme cases this condition is easily detected after endurance testing. Careful inspections should be made after such tests to detect borderline cases, as these can cause significant noise even if they are not severe enough to affect durability.

5.4.3 Vane-Intervane Pounding

Similar situations occur in vane pumps. If the undervane load is not sufficient to provide enough acceleration and overcome the viscous friction forces, the vane momentarily loses contact with the ring. This also occurs when rotor deflection exceeds the very small vane clearance and the vane is momentarily pinched. Later when proper conditions prevail, impact occurs as the ring contact is restored. This action contributes significantly to the total pump noise.

A similar impact occurs in pumps that use an intervane to supply needed undervane force. Whenever the pressures on their top and bottom are equal, centrifugal force propels them outward. If they are free to move, they impact the undervane surface. It is not clear how this produces noise, but in cases where it was eliminated, the pump became measurably quieter.

5.4.4 Gear Rattles

When pump gears mesh to form the seal between discharge and inlet, fluid trapped in tooth spaces can be compressed to very high pressures. This momentarily reverses the bearing loads and causes the gears to impact against the housing. This condition and its cure are discussed in Section 2.3.1. Fortunately, this problem tends to occur at low discharge pressures and is easily diagnosed by ear.

5.5 EFFECT OF STRUCTURE

The pump structure appears to offer few opportunities for reducing noise. Generally, functional requirements dictate that this structure be fairly beefy, so it is already both massive and stiff. Adding a little more metal or changing shape slightly, is therefore not likely to reduce greatly its response to pumping forces. This does not preclude the possibility of noise reduction by structure modification.

The effect of structural changes in cast iron parts is investigated experimentally by using brazing to add material. Braze material has about the same modulus of elasticity as cast iron, so modified parts behave about the same as cast iron parts with the same dimensions. Aluminum parts are similarly modified for test purposes, but this requires relatively sophisticated aluminum welding.

This cut and try approach is known only to have reduced the noise of one pump. It is interesting to note that the noise reduction was inversely proportional to the increase in the pump housing weight. At this rate the increased cost is seldom justified by such small benefit. The benefit being proportional to weight suggests that the noise generating system operated above its natural frequency, in its mass-controlled region. It is also concluded that the principal noise source did not involve plate bending, which is inversely proportional to the cube of wall thickness.

5.5.1 Structural Isolation

Concepts for providing vibration isolation within pump structures are often proposed. These would be effective in reducing the noise energy that reaches the outer radiation surfaces. However, no practical way of providing this isolation has been developed to date.

Isolation requires that structural stiffnesses be reduced so that the combination of stiffness and supported masses has a relatively low natural frequency. This is usually accomplished by inserting rubber-like or plastic materials in the load path within the structure. Unfortunately, the loads generally have high static components, which we ignored in most of our discussion because they do not cause noise. However when enough resilient material is provided to support this load, with reasonable durability, it is not very flexible. For this reason, isolation is generally reserved for use outside the pump, where only the torque is carried.

5.5.2 Damping

One method often called for in quieting pumps is to add damping. Often, this is intended as a generic term meaning to reduce vibration by any means. However, it specifically means that the conversion of vibratory energy to heat and people confuse the specific and generic terms.

Damping comes from several sources. Viscous damping arises from shearing forces in fluids and is proportional to velocity. The damper in the discussion of a simple vibration system in Section 3.1.2 was of this type. When materials are strained, internal friction causes damping that is often referred to as hysteresis. Minute slippage occurring in bolted joints or between hydraulically clamped surfaces also dissipates energy through friction. Damping from these two sources is proportional to strain or deflection. Although, mathematically, viscous and friction damping should be handled differently, they are so similar in effect that they are often lumped together and considered to be of either type.

Cast iron inherently has high-hysteresis damping. Aluminum generally has low hysteresis, but pumps with housings made of this material do not seem appreciably louder than comparable ones with cast iron housings.

The highest damping occurs in a special copper-manganese alloy. There is also a high damping alloy of magnesium. Attempts to have pump housing made of these two alloys failed because no foundry was found that would supply them. Plastics and some inorganic materials sold for vibration control also have high damping. Some of these can be laminated to load-carrying structural members and provide added damping.

There is one case known where a pump made of high-damping copper-manganese alloy was compared to one made of cast iron (6). It was reported that the pump vibration levels above 1000 Hz were attenuated by from 10 to 15 dB. An airborne noise comparison was not reported. This suggest that audible noise was not similarly reduced.

Damping is effective only in reducing vibrations near a resonant frequency or when impact causes free vibrations. In the latter case the maximum vibration remains essentially the same, but increased damping causes faster decay, so that the time average of the vibration is reduced.

The biggest drawback to trying to reduce pump noise by the use of high-damping materials is that this property is sensitive to temperature. All of these materials, which were considered for research studies, lose their high damping at common pump operating temperatures.

5.5.3 New Structural Analysis Tools

There is increased interest in reducing pump noise by improving their structure. One reason this has not been given a lot of consideration in the past is that it was very difficult to define the behavior of these structures. Efforts to develop improvements were usually limited to cut and try. A number of new developments are now available and are being tried (7,8).

One of these new tools is finite element analysis using a computer (7). The model of the structure is divided into small, connected elements having simple shapes. A computer then determines how these elements share loads and are deflected. Vibration modes and natural frequencies are calculated in this way. Critical areas are identified and the effects of modifications determined. The greatest value of this approach lies in the fact that these studies can be made in the design stage, before expensive patterns and tooling are bought. It is expected that this tool is effective in reducing not only noise but also cost.

The other tool is experimental modal analysis (7,8). This uses several techniques to determine the vibration modes of existing structures, their natural frequencies, and their damping. One method is to measure vibrations, with accelerometers mounted at a number of points on the structure, that result from a hammer blow. The hammer used for these tests has a transducer that measures the force profile of the blows. A computer compares the Fourier analysis of this impact with the resulting vibration and stores these data. Later the computer collates the data from the many measuring points and defines the vibration modes that occurred. This method is comparatively fast and is particularly useful for exploratory studies.

The other method of modal analysis uses an electromechanical shaker to cause vibrations. The shaker is powered with a random electrical signal so that the force supplied to the structure contains all frequencies of interest. Response data are collected over a relatively long time and averaged, so this method is more precise than the impact technique.

REFERENCES

1. NFPA T3.9.12, *Method of Measuring Sound Generated by Hydraulic Pumps*, NFPA Recommended Standard, National Fluid Power Association, Inc., Milwaukee, Wisc.
2. W. Ernst, *Oil Hydraulic Power and Its Industrial Applications*, 2nd ed., McGraw-Hill, New York, 1960.
3. R. Hadekel, *Displacement Pumps and Motors*, Sir Issac Pitman, London, 1951.
4. T. U. Thoma, *Modern Oilhydraulic Engineering*, Trade and Technical Press, Morden, Surrey, England, 1970.
5. W. E. Wilson, *Positive Displacement Pumps and Fluid Motors*, Pitman, Marshfield, Mass., 1950.
6. L. Kane, T. D. Richmond, and D. N. Robb, "Noise in Hydrostatic Systems and Its Suppression," *Proceedings Institution of Mechanical Engineers*, London, Part 3L, 180:1965–1966.
7. R. W. Dunlop, "Determination of the Natural Frequencies, Mode Shapes and Modal Damping Coefficients of a Positive Displacement Pump," *Proceedings of the Quiet Oil Hydraulic Systems Seminar*, Institution of Mechanical Engineers, Nov. 1977, pp. 139–154.
8. M. McKenna, "Experimental Structural Response of Hydraulic Components Using Digital Transfer Function Techniques," *Proceedings of the Quieter Oil Hydraulics Seminar*, Institution of Mechanical Engineers, Oct. 1980, pp. 25–33.

6
Fluidborne Noise

Our concern with fluidborne noise is that it has as much as a thousand times the energy of pump airborne noise. It therefore poses the risk of exciting other machine components to produce high airborne noise levels. Even when the conversion is relatively inefficient, it can produce louder sounds than the pump itself.

Pulsations generate vibrations as well as sound. In aircraft, for example, they are responsible for hydraulic line vibrations that cause metal lines to fail in fatigue and flexible lines to fail by chaffing. Hydraulic elevator manufacturers are concerned about pulsations because they cause passenger discomfort. Often, these fluidborne noise-induced vibrations, in turn, also generate sound.

Fluidborne noise is generally more important in mobile machinery than in industrial. From my experience in quieting industrial machines, there has been only one case where fluidborne noise was the principal cause of trouble. However, in mobile equipment where hydraulic lines are often an integral part of structural members, fluidborne noise has a very high potential for causing noise problems. In the case of a large bucket-type scraper that had its hydraulic lines welded to the bucket and an experimental pump that generated very high levels of fluidborne noise, the airborne hydraulic noise was clearly discernible over the large diesel engine's noise.

There is a new and growing concern over fluidborne noise which is beyond the scope of this book. Pulsations can interfere with the electronic controls that are finding wider use in hydraulic systems. Most of these depend on pressure measurements to govern pump output. The presence of pressure oscillations generally requires more sophisticated electronics for signal processing and may also limit the system's sensitivity.

6.1 FLUIDBORNE NOISE MECHANICS

Fluidborne noise is periodic fluid flow perturbations or pulsations. When these encounter flow resistance, they result in pressure pulsations as well. Generally, our major concern is the effect of the pressure disturbances. For this reason we generally consider fluidborne noise to be pressure pulsations or waves.

6.1.1 Some Basics

The major difference in airborne and fluidborne noise is the medium in which they exist. Because fluids have higher stiffness or bulk moduli, fluid waves travel faster than air waves. The sound velocity in a fluid is

$$c = \sqrt{\frac{B}{\rho}} \quad \text{in./sec}$$

where

B = bulk modulus, psi

ρ = mass density, $\text{lb-sec}^2/\text{in.}^4$

It follows that the length of fluidborne waves, the distance between like points on successive waves, is

$$\text{wavelength, } \lambda = \frac{c}{f} \text{ in.}$$

where

c = sonic velocity, in./sec

f = frequency, Hz

It is often necessary to deal with fractions of a wavelength. In this we consider wavelength in terms of angles, with one wavelength equal to 2π rad. Angular fractions of a wavelength are found by multiplying distances by a coefficient equal to

$$\beta = \frac{2\pi}{\lambda} \text{ rad/in.}$$

This quantity is called the *phase shift constant* because the phase difference in pressures at two points is determined by multiplying it by the distance between the points.

6.1.2 Bulk Modulus

It is extremely difficult to find suitable values for bulk modulus in the literature. The reason is that it varies with measurement method, pressure, and temperature. Care must be taken not to use an isothermal bulk modulus since this was determined for slow incremental changes in pressure where the temperature remained constant. The adiabatic, isentropic, or dynamic bulk modulus which is determined with fast pressure changes is the appropriate one to use. This is from 10 to 20% higher than the isothermal. Values are also given for small pressure ranges (tangent) or the average over large pressure ranges (secant). This diversity, coupled with the variations due to pressure and temperature, make the selection very difficult.

After sorting through all these parameters it was found that published values did not always agree. For this reason I resort to homemade data. A colleague at Vickers, Ron Becker, determined the following velocities and bulk moduli using an ultrasonic device. These measurements were made at room temperature and atmospheric pressure. Although they lack the sophistication of other data given in the literature, they are good for comparisons between fluids.

FLUID PROPERTIES

Fluid	Mass density ($lb\text{-}sec^2/in.^4$)	Bulk modulus (psi)	Sonic velocity (in. /sec)
Petroleum oil	9.35 x 10	316.000	58,000
High water base	8.17 x 10	259,000	56,300
MIL-O-5606	8.13 x 10	232.000	53,500

In hydraulic circuits the effective bulk modulus, as determined by wavelength measurements, is almost always much lower than that measured in fluid laboratories. The same was true when pressure profiles were measured and compared with calculated profiles. Toet (1) found that the wave velocity did not necessarily remain constant throughout a test and may be as much as 30% below predictions from fluid data. Stevens (2) also reported a wave velocity that is over 20% lower than calculated. Entrained oil appears to be the reason for the variance. It requires only ½% of entrained air, at 2000 psi, to reduce the wave velocity by 30%.

Correlation with fluids data is possible. Considerable effort was made to purge air from the fluid in an acoustic filter test circuit. In addition, the static pressure and temperature were constant throughout this system, which also had no flow. Measured wave velocities in this test rig were very close to calculated values (3). The results also show that the speed of sound varies with frequency.

For some time, the author has successfully used a bulk modulus of about 180,000 psi and a velocity of about 47,000 ips for common circuits. These are averages derived from measurements in pumps and hydraulic systems using petroleum oil. This result also agrees with ultrasonically measured data (4) and the pump quieting experience reported by Louthan (5). If we use this value as a base, working values for the other fluids are found by reducing proportionally the values in the table.

6.1.3 Hydraulic Circuit Analysis

There is a practical difference between airborne and fluidborne noise. Fluidborne noise is generally confined to conduits whose diameter is small in comparison to the wavelength. It therefore propagates as plane waves, which are simpler to analyze than the spherical waves that are normal for sound.

The reaction of fluidborne noise to hydraulic acoustic filters and simple circuits is analyzed by the impedance method (6−8). As in the analysis of other vibrations, fluidborne noise is analyzed for one frequence at a time.

The impedance at a point in a circuit is defined as

$$Z = \frac{P}{Q} \text{ lb sec/in.}^5$$

where

P = noise pressure amplitude at the point, psi (rms)

Q = noise volume flow amplitude at the point, in.^3/sec (rms)

Impedance in series, as shown in Figure 6.1(a), are simply added. The impedance at A in this figure is

$$Z_A = Z_1 + Z_3 + Z_4$$

When an impedance is in a branch, sometimes called a shunt, as Z_2 in Figure 6.1(b), flow is divided between it and the main circuit in the inverse proportion to their impedances. Their combined impedance is the sum of their reciprocals. The impedance at A in Figure 6.1(b) is

$$Z_A = Z_1 + \frac{1}{\frac{1}{Z_2} + \frac{1}{Z_3 + Z_4}}$$

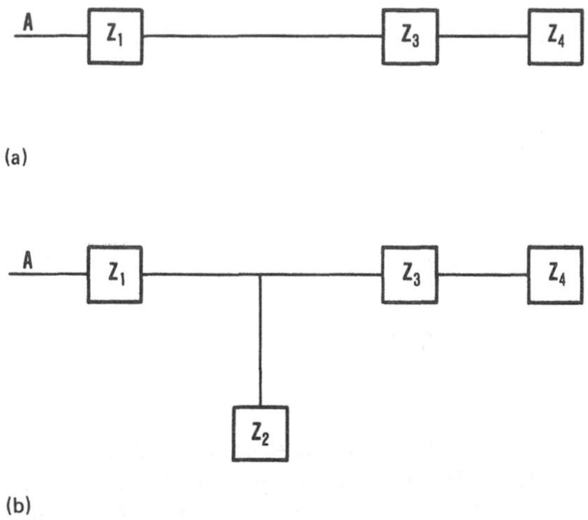

FIGURE 6.1 (a) Series impedances; (b) Shunt impedance.

Impedances are vectors and can be complex numbers. When a term includes the letter j, it indicates that this term is a vector that is at right angles to terms that do not include a j. This j (sometimes the letter i is used) is the algebraic symbol for the square root of -1. Terms including it are referred to as *imaginary numbers;* those without it are called *real numbers*. Real and imaginary numbers can be added only by vector addition. Arithmetically, this is accomplished by adding the squares of both terms and taking the square root of this sum. In doing this the negative derived from squaring j is ignored and the absolute values of the squares are summed. All imaginary terms, of course, can be added to each other arithmetically, as can all real terms.

Some people prefer to employ linear circuit theory (9), as used in electrical engineering, to analyze fluid pulsations in hydraulic circuits. This is done by assuming that voltage is analogous to fluid pressure and current is analogous to volume flow (in.^{3}/sec). For reference, other electrical analogies are given as we discuss impedances of various hydraulic elements. The author is not qualified to cover electrical theory, so reference to appropriate textbooks or practitioners of the art is recommended.

The compressibility of fluid volumes is analogous to capacitance in electrical terms and spring rate in mechanical. For a volume whose largest dimension is small in comparison to the wavelength of the pressure wave, the impedance is

$$Z_c = -\frac{j\rho c^2}{V\omega} \text{ lb-sec/in.}^5$$

where

j = $\sqrt{-1}$ used to indicate 90° phase difference

ρ = density, lb-sec^2/in.^{4}

c = acoustic velocity, in./sec

v = fluid volume, in.^{3}

ω = circular frequency, $2\pi f$, rad/sec

f = frequency, Hz

When a short tube is connected to a fluid volume of the type just discussed, the fluid mass in the tube oscillates back and forth with the pulsations. It therefore provides inertia whose electrical analogy is inductance. The impedance of this fluid mass is

$$Z_I = \frac{j\rho l\omega}{A} \text{ lb-sec/in.}^5$$

where

l = tube length, in.

A = tube cross-sectional area, in.^{2}

Generally, we must consider both compressibility and mass of fluid in longer tubes or lines. This is realized in the *characteristic impedance* of a hydraulic line

$$Z_o = \frac{\rho c}{A} \text{ lb-sec/in.}^5$$

Valves controlling flow in hydraulic circuits have an impedance of

$$Z_T = \frac{n\bar{P}}{\bar{Q}} \text{ lb-sec/in.}^5$$

where

$\bar{P}$ = average pressure drop across valve, psi

$\bar{Q}$ = average volume flow through valve, in.^{3}/sec

n = coefficient depending on type of flow

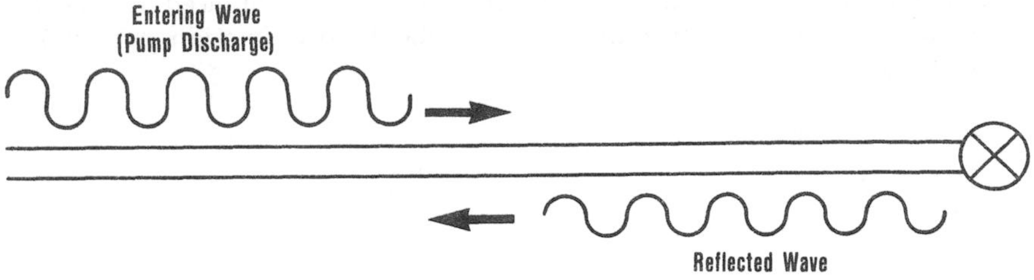

FIGURE 6.2 Pressure waves in a pump discharge line.

Average here means gage pressure and the steady flow, not the average of the pulsations, which is zero. Theoretically, the coefficient in the equation above should be 2 if the flow through the valve is turbulent and 1 if it is laminar. It is usually somewhere between 1.5 to 2 for most valves and may also vary with frequency (9).

6.1.4 Standing Waves

Analysis of fluidborne noise in a simple pump discharge line terminated by a valve, like that shown in Figure 6.2, provides a good picture of how pressure waves react in hydraulic circuits. The pump supplies a periodic noise flow superimposed on the steady flow. The instantaneous volume velocity of one harmonic component of this noise flow is

$$q = Av \sin \omega t \quad \text{in.}^3/\text{sec}$$

where

A = cross-sectional area of line, in.^{2}

v = maximum noise flow velocity, in./sec

ω = circular frequency of noise, Hz

t = time, sec

The root-mean-square value of this flow is then

$$Q = 0.707Av \quad \text{in.}^3/\text{sec}$$

If the line is so long that the pressure wave is dissipated before it reached the end, the pressure in the tube before friction reduced it would be

$$P = Z_0 Q \quad \text{psi}$$

where

Z_0 is the line characteristic impedance in lb-sec/in.^{5}

The line shown in Figure 6.2 is not that long and friction will not diminish the wave appreciably before it reaches the valve. If the valve impedance is equal to the line characteristic impedance, the entire wave will pass through the valve. Usually, the impedances are not equal and only a fraction of the wave passes through the valve, with the rest reflected back down the line toward the pump. The fraction reflected is

$$r = \frac{Z_T - Z_0}{Z_T + Z_0}$$

where

Z_T = valve impedance, lb-sec/in.^{5}

Z_0 = line characteristic impedance, lb-sec/in.^{5}

The pressure at any point in the line is the sum of the original and reflected waves. With a line terminated by a valve, the phase angle of the reflected wave is not changed, so the waves add arithmetically at the valve. This is true if the valve has purely resistive impedance. If the impedance includes some inertia or compressive effects, there is a phase shift as the wave is reflected. In any case, pressure at the valve is

$$P_T = \frac{Q_\varepsilon Z_0 Z_T}{Z_0 \cos \beta l + jZ_T \sin \beta l} \quad \text{psi (rms)}$$

where

Q_ε = periodic volume velocity entering line, in.^{3}/sec (rms)

Z_0 = line characteristic impedance, lb-sec/in.^{5}

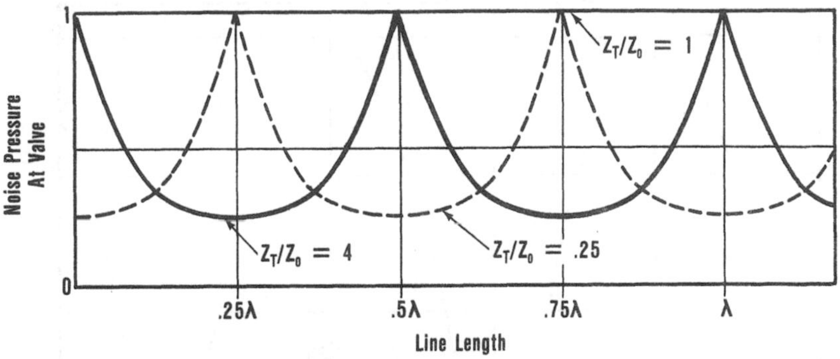

FIGURE 6.3 Effect of discharge line length on noise pressure at valve.

Z_T = valve impedance, lb-sec/in.^{5}

l = line length, in.

β = phase-shift constant, rad/in.

It can be seen that the pressure at the valve not only depends on the line and valve impedances but also on the line length. Organ pipe resonances occur in these lines, as seen in Figure 6.3. If Z_T is larger than Z_0, resonance occurs and noise pressure at the valve is highest when the line length is a multiple of a half-wavelength ($\lambda/2$, λ, $3\lambda/2$, ...). Pressure will be lowest when the line length is an odd multiple of a quarter-wavelength ($\lambda/4$, $3\lambda/4$, $5\lambda/4$, ...). When Z_0 is the larger, the highest pressures occur with line lengths that are odd multiples of a quarter-wavelength and the lowest with multiples of a half-wavelength.

When the original wave passes through a point, it must travel to the valve and then back through this same distance as the reflected wave. If the valve impedance is purely resistive, the phase angle between the two waves at this point then is

$$\varnothing = 2x\beta \text{ rad}$$

where

β = phase shift constant, rad/in.

x = distance measured from valve, in.

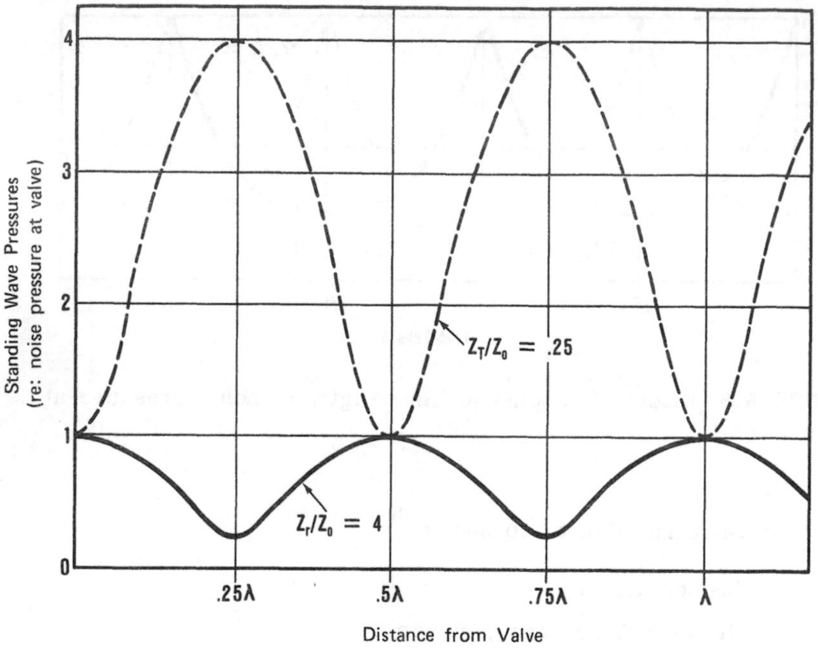

FIGURE 6.4 Noise pressures along discharge line.

Since the pressure at the point is the vector sum of the two waves, this sum varies with the phase difference in the waves. This angle is a function of the distance from the valve, so the total pressure at the point also varies with the distance from the valve. The pressure at any point is

$$P = \frac{P_T(Z_T \cos \beta X + jZ_0 \sin \beta X)}{Z_T} \text{ psi (rms)}$$

This equation is general and is valid even when the line terminal impedance is not just resistance.

The shape of this function is wavelike, as shown in Figure 6.4, and is called a *standing wave.* If the line is more than a quarter-wavelength long, the ratio of the highest to the lowest pressure is called the *standing-wave ratio* and is sometimes used as a measure of the strength of the reflection from the valve.

While this pressure profile remains stationary, the pressure at any point is still periodic, the instantaneous pressure at any point being

$$p = 1.4P \sin \omega t \quad \text{psi}$$

where

ω = circular frequency of waves, rad/sec

P = standing-wave pressure, psi (rms)

6.1.5 Computer Simulation

The impedance method is useful in understanding the behavior of fluidborne noise in simple circuits and for identifying possible fluid resonances. This is all that may be needed for practical noise reduction work. In rare cases it may be desirable to analyze a complex circuit's reaction to fluidborne noise. Computer analysis is then preferred.

McDonnell Douglas Aircraft Corp. has written a comprehensive program that analyzes complex hydraulic systems like those found in aircraft. This work was done for the U.S. Air Force. Magnetic tape copies are available for many mainframe computers along with a user's manual (10).

The program is called HSFR (Hydraulic System Frequency Response). Circuit component parameters, in physical terms, are entered into the computer in accordance with instructions in the user's manual. Pump speed, system pressure, temperature, and flow are also entered. The program then computes acoustic flows, noise pressures, and impedances at selected points in the system, for selected harmonics.

Unfortunately, the program models the pump, and this model is not suitable for other pump types. However, the program is written in Fortran, so can be modified to provide a more suitable pump model or, better yet, just a unit noise flow, so that the results are normalized.

Program results for a simple laboratory test circuit were compared with those measured in the circuit. Agreement was fair for the first two harmonics but unacceptable for higher harmonics. Unfortunately, no follow-up work to identify the source of the discrepancies was attempted.

Hydraulic system noise analyses are also made with computer programs for solving differential equations. These are generally used for analyzing hydraulic control responses. Although these require writing a number of differential equations to describe the system components, this process has been simplified and reduced almost to a cookbook procedure.

Two of these programs have been recommended to me for this purpose. The first is called MIMIC (11). It was developed at the Wright-Patterson Air Force Base and has been released for public use.

The second program is newer and is said to be more versatile. It is called ACSL (12) (pronounced "axle"; Advanced Continuous Simulation Language).

6.2 PUMP FLUIDBORNE NOISE

Pumps and motors are the leading source of fluidborne noise. Although not nearly as important a source, valves also produce this noise, and this is a subject of Chapter 8.

An oscilloscope trace of pressure pulsations in a pump discharge is shown in Figure 6.5. A typical pump fluidborne noise spectrum, not related to that shown in the preceding figure, is shown in Figure 6.6. It differs from those of airborne noise in that there are peaks at the shaft frequency and several of its harmonics. Also, the pumping fundamental is higher than its harmonics. Both of these features are typical and are due to the absence of frequency weighting like that of airborne noise.

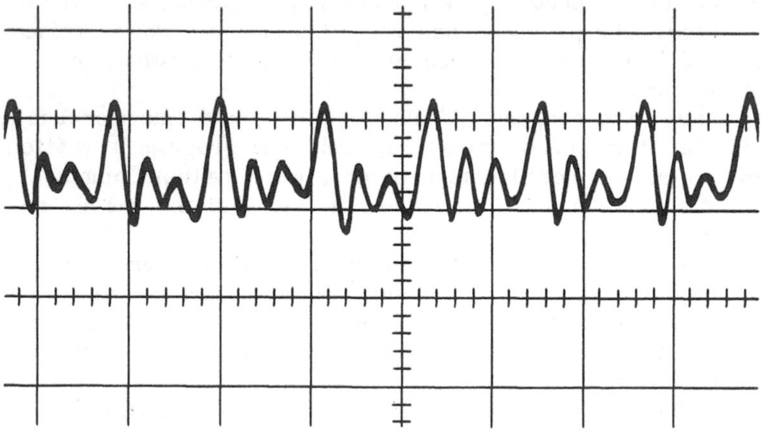

FIGURE 6.5 Oscilloscope trace of typical pump discharge pressure pulsations.

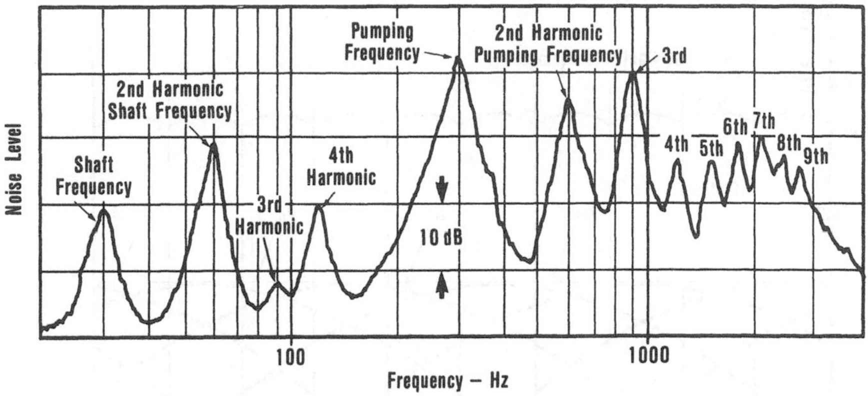

FIGURE 6.6 Typical fluidborne noise spectrum, 10-vane pump.

6.2.1 Pump as a Source

Positive-displacement pump fluidborne noise is due primarily to the discrete nature of the pumping process and to fluid compressibility. Some periodic flow is also due to pumping geometry, cavitation, machining imperfections, and even small variations in the drive speed. These sources generally are so minor that their presence is difficult to detect in analyzing fluidborne noise spectra.

Pumps divide the incoming fluid stream into a series of pumping chamber volumes, compress these volumes to raise their pressure, and then recombine the volumes into a discharge flow. Some of the mechanics of this process are discussed in earlier chapters. In particular, Section 5.2.3, dealing with the pressurization process, is important to our discussion here. This process is best seen by examining the action of an inline piston pump.

Inherently, the discharge volume from one cylinder is a half-sine function of shaft rotation. The theoretical pump discharge, is then the sum of these equally spaced half-sines. Figure 6.7 shows this sum for a nine-cylinder pump.

This combining of half-sines is the same as that done in the discussion of shaft torque in Section 2.1.2. The resulting ripple or fluidborne noise, for odd numbers of cylinder pumps, is equal to the function W, given in that section, and function W' for even numbers of cylinders. Figure 6.8 shows how the amplitude of this ripple varies with the number of cylinders. The fundamental frequency of this ripple, for odd numbers of cylinders, is twice piston frequency, and that for even numbers is equal to piston frequency.

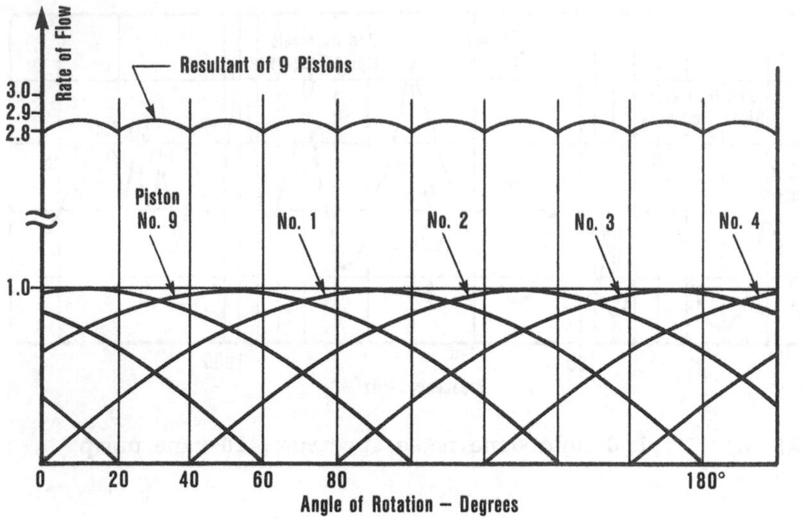

FIGURE 6.7 Inherent piston pump discharge ripple.

This is the inherent or geometric fluidborne noise of an inline piston pump, and it occurs even when the pump is not increasing fluid pressure.

If the pump has "line-to-line" porting, it connects a cylinder to the discharge port and disconnects it from the inlet port at the instant that it reaches bottom dead center. The fluid in the cylinder is still at inlet pressure when it is brought into contact with discharge pressure. The fluid in the discharge port therefore flows backward into the cylinder to compress the fluid to the discharge pressure. This backflow periodically reduces the pump discharge as shown in Figure 6.9. The flow perturbations due to this action, which are much greater than the inherent ripple, are the major fluidborne noise component.

6.2.2 Effect of Port Timing

Line-to-line porting causes the highest fluidborne noise levels possible with a given pump. However, no matter what type of porting is provided, the need to compress the pumping chamber fluid will always cause some fluidborne noise. In general, extending the time for the compression process reduces the noise level. As discussed

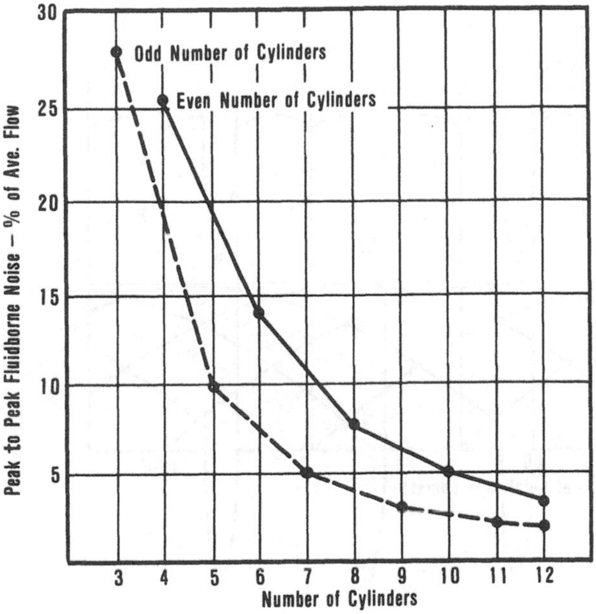

FIGURE 6.8 Fluidborne noise due to flow ripple inherent in piston pumps.

in Section 5.2.3, the maximum increase in this time, with simple porting, is achieved by delaying the opening to discharge until piston motion provides the required compression. Even this strategy causes appreciable fluidborne noise because cylinders are not providing their share of discharge flow during this compression period. The effect is as shown in Figure 6.10.

As with airborne noise, delayed porting provides only minimum noise for the discharge pressure provided by the port opening delay. Portings that use a combination of metering groove and delayed port opening cannot achieve the low noise levels possible with delayed porting alone. However, they provide appreciably reduced levels over a wider range of pressures. Discharge flow with metering grooves and delayed porting is illustrated in Figure 6.11. It can be seen that portings that reduce airborne noise reduce fluidborne noise as well.

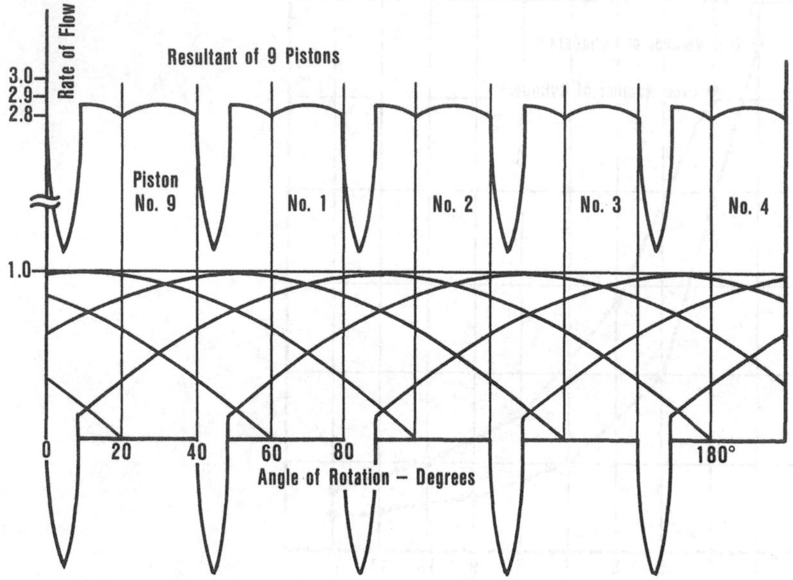

FIGURE 6.9 Fluidborne noise with line-to-line porting.

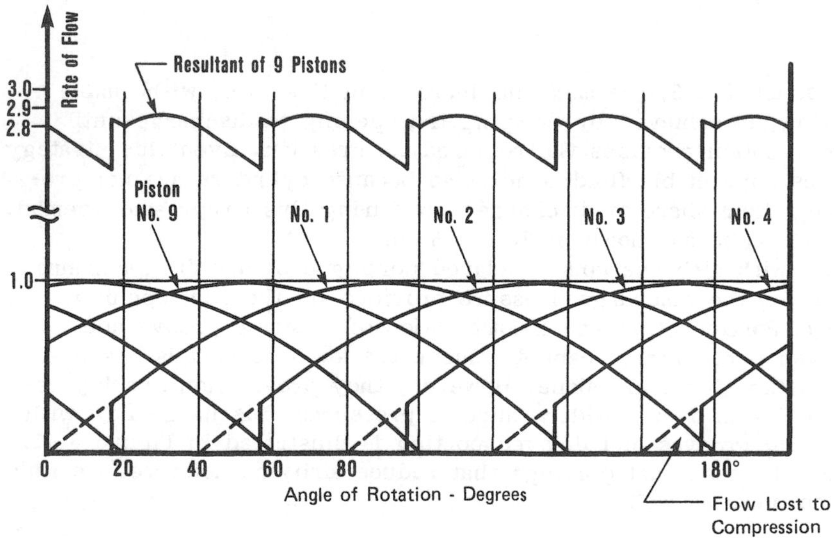

FIGURE 6.10 Fluidborne noise with delayed porting.

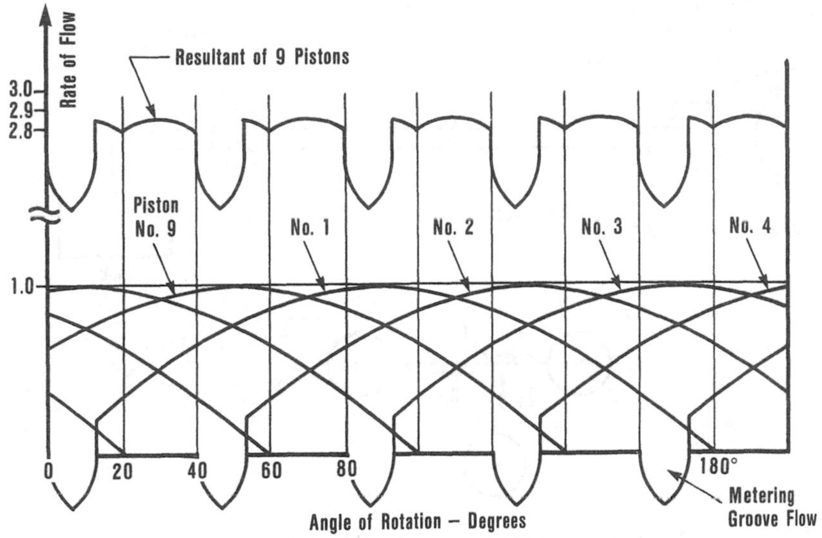

FIGURE 6.11 Fluidborne noise with delayed porting and a metering groove.

6.2.3 Pump Fluidborne Noise Standards

Considerable effort was expended in developing a fluidborne noise standard. The most comprehensive effort was made in Great Britain, under government sponsorship. In particular, the work performed at the University of Bath by D. E. Bowns and his associates (6,9, 13–15) has been the most effective in the development of a standard.

In the United States the leading work has been done at the David W. Taylor Naval Ship Research and Development center (7), with hydraulic equipment manufacturers and other laboratories also contributing (2,16–19). Studies are also known to have been made in the Netherlands (1) and France.

The University of Bath research resulted in a British standard for measuring pump fluidborne noise (20). This document has also been submitted to the International Standards Organization (ISO) as an international standard proposal (21). Because of the difficulties in developing an international standard and securing the approval of the many interested countries, it may be the end of this decade before an international standard is adopted.

Work in the United States has not yet produced a standard. Draft standards follow the principal concept of the British standard

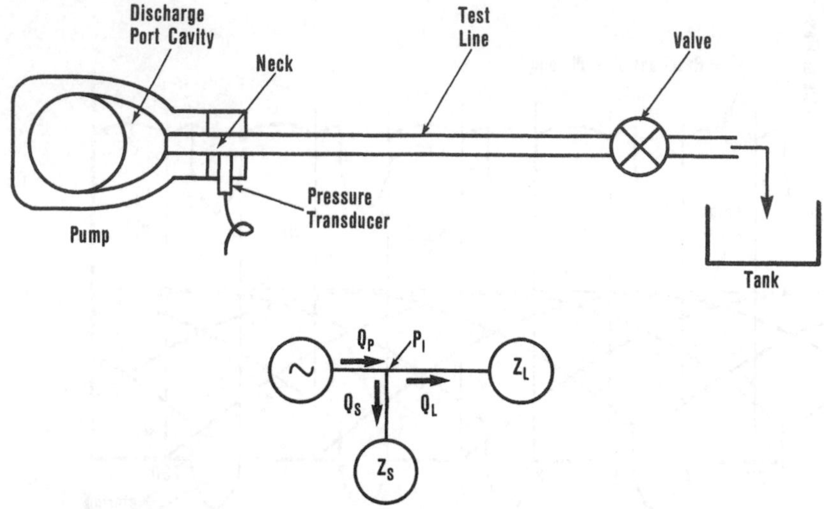

FIGURE 6.12 Schematic diagrams of pump test circuit.

but seek to reduce the experimental work required to evaluate a pump. Because of the press of other work, progress is slow. It is possible that the international standard may be available before an American one is adopted.

6.2.4 Measuring Pump Fluidborne Noise

In Section 6.1.4 we showed how pump pressure pulsations are influenced by discharge-line parameters. Because the discharge line has this much influence, there is a problem in finding a measurement that accurately scales a pump's intrinsic fluidborne noise without being affected by the test circuit.

The hydraulic circuit discussion in Section 6.1.3 did not include the influence of pump internal impedance on fluidborne noise. The schematic diagram of a pump and test circuit in Figure 6.12 shows how this impedance shares pump noise flow with the test circuit impedance. A cavity in the pump represents the fluid volume in the discharge passages. This volume is the primary factor in determining the pump impedance.

The periodic flow generated by the pump, at each frequency, is divided between the pump and test circuit impedances in inverse proportion to their magnitudes. This flow then is

$$Q_P = P_I \left[\frac{1}{Z_L} + \frac{1}{Z_S}\right] \text{ in.}^3\text{/sec (rms)}$$

where

P_I = measured pressure at interface, psi (rms)

Z_S = pump internal impedance, lb-sec/in.^{5}

Z_L = test circuit impedance at interface, lb-sec/in.^{5}

Solving this for the pressure measured at the interface gives

$$P_I = \frac{Q_P Z_S}{1 + Z_S/Z_L} \quad \text{psi (rms)}$$

The numerator consists of two inherent pump parameters. Their product is called the blocked pressure because it is the pressure pulsation that would be generated if the pump only produced its noise flow and its outlet was blocked. It is an inherent property of the pump that is analogous to the open-circuit voltage of an electrical generator.

Although pump flow perturbations are the basic cause of fluidborne noise, it is generally agreed that the blocked pressure is the best measure of this noise. The principal reason is that dynamic pressures are easily measured, whereas dynamic flows are very difficult to measure.

The blocked pressure is measured by the transducer at the pump interface if the test circuit impedance is much higher than the pump impedance. This is seen from the preceding equation, where this condition would cause the denominator to approach 1. When the circuit impedance is eight or more times greater than the pumps, the measured pressure is within 1 dB of the actual blocked pressure.

The impedance of the test circuit is

$$Z_L = Z_0 \frac{Z_T \cos \beta l + jZ_0 \sin \beta l}{Z_0 \cos \beta l + jZ_T \sin \beta l} \quad \text{lb-sec/in.}^5$$

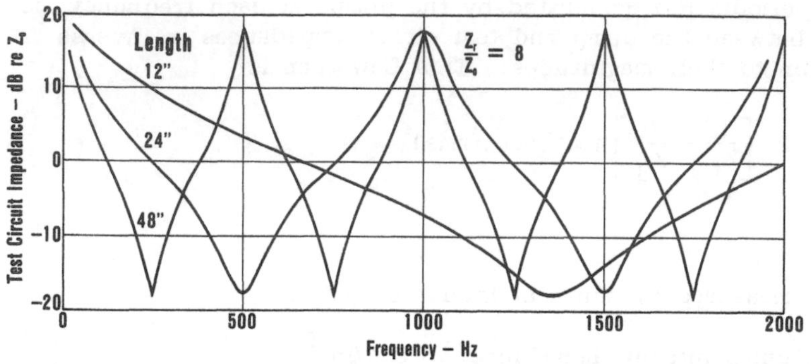

FIGURE 6.13 Effect of discharge line length on circuit impedance.

where

Z_T = load valve impedance, lb-sec/in^5.

Z_0 = test-line characteristic impedance, lb-sec/in^5.

β = phase shift constant, rad/in.

l = test-line length, in.

Figure 6.13 shows how this impedance, for three different line lengths, varies with frequency.

The trigonometric terms in the equation above cause the impedance to go through a complete cycle whenever the term βl varies through a range of π, which corresponds to one-half wavelength. The ratio of the valve impedance and line characteristic impedance determines the amplitude of this cycling, as seen in Figure 6.14.

Since the trigonometric terms only cause cycling, the only way to raise the test circuit impedance at all frequencies is to increase the line characteristic impedance Z_0, which is the only factor outside these terms. This is done by using a smaller-diameter line, which as shown in Section 6.1.3, increases impedance inversely with the square of the diameter. The British standard recommends that the line size be reduced until the pressure drop in the test line equals about three-fourths the discharge pressure.

In most pumps, the compressibility of the fluid in the discharge passages and the pumping chambers in communication with the discharge port account for most of the pump impedance. This impedance, which is also given in Section 6.1.3, is

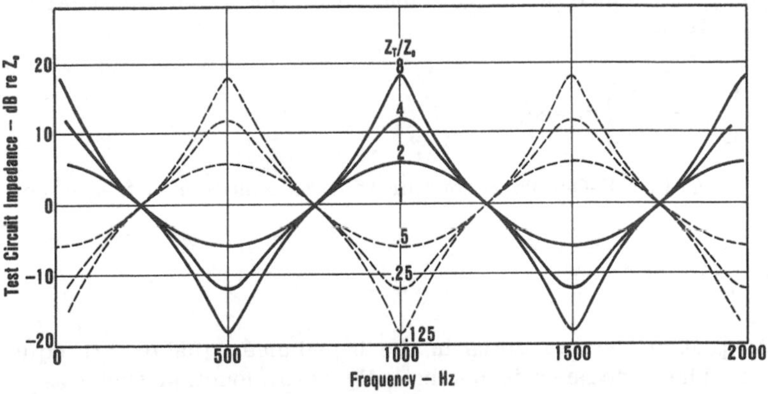

FIGURE 6.14 Effect of valve impedance on circuit impedance.

$$Z_c = -\frac{j\rho c^2}{V\omega} \text{ lb-sec/in.}^5$$

where

ρ = fluid density, lb-sec^2/in.^{4}

c = acoustic velocity, in./sec

v = fluid volume, in.^{3}

ω = circular frequency, rad/sec

Discharge passages generally neck down at the outlet port. Fluid in this neck provides inertia having an impedance of

$$Z_I = \frac{j\rho l\omega}{A} \text{ lb-sec/in.}^5$$

where

l = neck length, in.

A = neck cross-sectional area, in.^{2}

The outlet geometry is not as neat as these equations imply. However, the neck effect is real and reasonable estimates of the parameters are possible. To conform with the test circuit model, the neck is considered to extend to the pressure-measuring transducer, even though this is beyond the outlet interface.

The pump impedance is the combination of these two elemental impedances. It is

$$Z_s = j\rho \left[\frac{l\omega}{A} - \frac{c^2}{V\omega} \right] \text{lb-sec/in.}^5$$

This subsystem is a Helmholz resonator that resonates at a frequency of

$$\omega_L = c \sqrt{\frac{A}{Vl}} \ \text{rad/sec}$$

At this frequency the two terms in the impedance equation are equal and cancel. The impedance is not actually zero. Friction that was omitted from the equation because it is unimportant at most other frequencies absorbs the energy. Interestingly, this resonance occurred at about 1000 Hz for most of the pumps tested at Naval Ship R and D Center, indicating a similarity in pump discharge port design. A resonance at this frequency is particularly important because it makes pump impedance low in the critical airborne noise region from 500 to 2000 Hz. It is therefore easier to provide discharge lines with relatively high impedances for accurate noise measurements in this range.

The impedance for a vane pump having a discharge port volume of 13.3 in.^{3} with a neck 0.75 in. in diameter and 3.338 in. long is shown in Figure 6.15. This curve was normalized by dividing by the characteristic impedance of the pump's standard (0.75 in.) discharge line.

The pump has a calculated resonant frequency of 750 Hz. At frequencies above resonance, the inertia term in the equation predominates, so the pump impedance rises with increasing frequency. The details of this portion of the curve are less certain because at high frequencies pump passages begin to be significant fractions of a wavelength and the impedance is affected by standing waves.

The impedance of a 0.75-in.-diamenter line 48 in. long is also shown in Figure 6.15. This curve too, is normalized. The error in assuming that the measured pressure is the blocked pressure is also shown below. It can be seen that accurate blocked pressure measurements can be made only over limited ranges of frequencies.

Following the strategy of the British standard, a 0.25-in.-diameter line would be used for this pump if it were operated at 2000 rev/min, delivering 24 gal/min at 2000 psi. Length in accordance with the standard is about 1 m (39.4 in.). Impedances for this line and the pump are shown in Figure 6.16. These are normalized to the higher characteristic impedance of the smaller line, so the pump impedance is displaced downward.

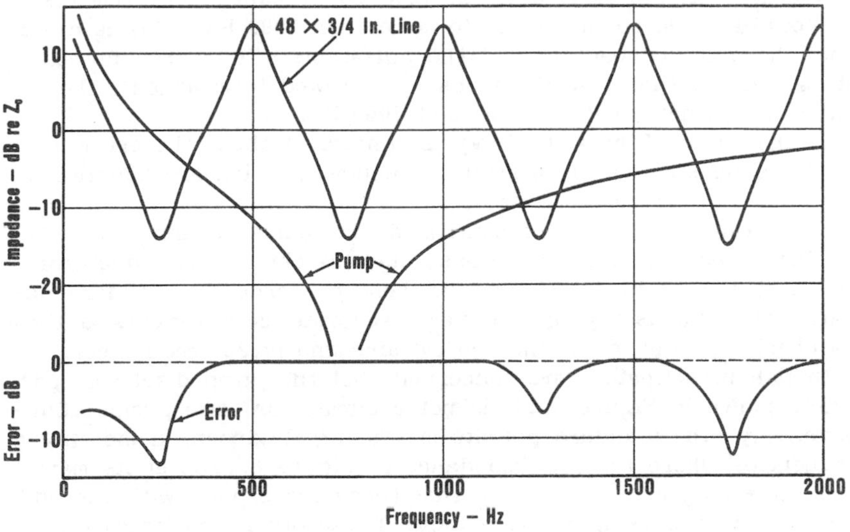

FIGURE 6.15 Standard line test impedances and errors.

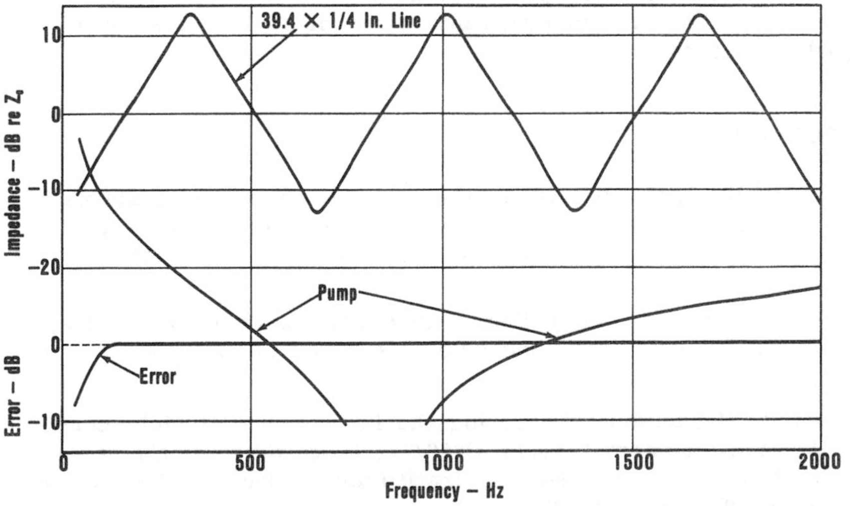

FIGURE 6.16 Small diameter line test impedances and error.

With this combination, accurate blocked pressure measurements are possible from about 100 Hz to more than 2000 Hz. This is quite good where the concern is that the pulsations will be transformed into sound. In that case the focus is narrowed to measuring the critical frequencies between 500 and 2000 Hz.

In the United States the Navy is concerned about the error at low frequencies. Pump rotational frequencies, which are apparently converted into seaborne noise as readily as higher frequencies, fall into this range and must be controlled. Companies dealing with fluidborne noise-excited vibrations, such as elevator manufacturers, have a similar problem. Because of these concerns the standard development group is trying to develop accurate measurements at these frequencies as well as at the critical airborne noise frequencies.

British investigators are concerned that the pump discharge port model, shown in Figure 6.16, is not accurate for some pumps. Conservatively, the standard permits measuring fluidborne noise only at frequencies where the line impedance in Figure 6.16 is at its maximum. Line length is selected so that these correspond with the odd harmonics of the pumping frequency. Three other line lengths are specified for measuring the remaining even harmonics.

The British Standard requires measuring the first 10 pumping frequency harmonics. Besides providing these levels, it provides a *pressure ripple rating*. This is found by taking the square root of the sum of the squares of the 10 harmonic pressures.

The standard uses the bar as the unit for pressure. Originally, this was defined as atmospheric pressure, which was generally taken as 14.7 psi. More recent documents define it as 10 Pa, which is equal to 14.5 psi.

REFERENCES

1. G. Toet, "Determination of the Vibration Behavior in Simple Hydraulic Systems," *Proceedings of the Fluid Power Testing Symposium*, Milwaukee School of Engineering, Aug. 1975, pp. 4.2.1–4.2.18.
2. F. A. Stevens, "Pump Flow Relationship to Fluidborne Noise," *Proceedings of the National Conference on Fluid Power*, Oct. 1977.
3. A. R. Henderson, "Measuring the Performance of Fluid-Borne Noise Attenuators," paper C256/77, *Proceedings of the Quiet Oil Hydraulic Systems Seminar*, Institution of Mechanical Engineers, Nov. 1977, pp. 15–28.
4. J. W. Noonan, "Ultrasonic Determination of the Bulk Modulus of Hydraulic Fluids," *Materials Research and Standards* 5: 615–621, Dec. 1965.

5. G. W. Louthan, "Hydraulic Noise Reduction by Piston Pump Modification," *Product Engineering*, pp 170–182, Feb. 1952.

6. D. E. Bowns and D. McCandlish, "Pressure Ripple Propagation," paper C264/77, *Proceedings of the Quiet Oil Hydraulic Systems Seminar,* Institution of Mechanical Engineers, Nov. 1977, pp. 93–102.

7. L. C. Davidson, "The Internal Impedance of Positive Displacement Pumps: Experimental Determination and Effect on System Noise," *Proceedings of the National Conference on Fluid Power,* 30: 35–37, Oct. 1976.

8. R. C. Binder, *Advanced Fluid Mechanics,* Vol. 2, Prentice-Hall, Englewood Cliffs, N.J. 1958, pp. 251–262.

9. D. McCandlish, K. A. Edge, and D. G. Tilley, "Fluid Borne Noise Generated by Positive Displacement Pumps," paper C265/77, *Proceedings of the Quiet Oil Hydraulic System Seminar,* Institution of Mechanical Engineers, Nov. 1977, pp. 103–114.

10. *HSFR Users Manual,* Technical Report AFAPL-TR-76-43, Vol. 4, Systems Engineering Group, Aeronautical Systems Division, Air Force Systems Command, Wright-Patterson Air Force Base, Ohio, Feb. 1971.

11. F. J. Sanson and H. E. Petersen, *MIMIC Program Manual,* Technical Report SEG-TR-67-31, Systems Engineering Group, Aeronautical systems Division, Air Force Systems Command, Wright-Patterson Air Force Base, Ohio, July 1967.

12. *Advanced Continuous Simulation Language (ACSL),* Mitchell and Gauthier, Associates, Inc., 290 Baker Avenue, Concord, Mass.

13. D. E. Bowns, K. A. Edge, and D. G. Tilley, "The Assessment of Pump Fluid Borne Noise," paper C266/77, *Proceedings of the Quiet Oil Hydraulic Systems Seminar,* Institution of Mechanical Engineers, Nov. 1977, pp. 115–125.

14. D. E. Bowns, K. A. Edge, and D. McCandlish, "Factors Affecting the Choice of a Standard Method for the Determination of Pump Pressure Ripple," paper C373/80, *Proceedings of the Quieter Oil Hydraulics Seminar,* Institution of Mechanical Engineers, Oct. 1980, pp. 1–6.

15. M. D. Butler and D. G. Tilley, "The Generation and Transmission of Fluid Borne Pressure Ripple in Hydraulic Systems," paper C374/80, *Proceedings of the Quieter Oil Hydraulics Seminar,* Institution of Mechanical Engineers, Oct. 1980, pp. 7–14.

16. S. F. Szerlag, "Rating Pump Fluidborne Noise," SAE paper 750830, Society of Automotive Engineers, Sept. 1975.

17. G. J. Czarnecki, "Fluidborne Noise Measurement and Analysis," *SESA Spring Meeting*, Society for Experimental Stress Analysis (now the Society for Experimental Mechanics), May 1976.

18. D. Unruh, "Outlet Pressure Ripple Measurement of Positive Displacement Hydraulic Pumps," *Proceedings of the National Conference on Fluid Power*, 29: 727–762, Oct. 1975.

19. D. L. O'Neal and G. E. Maroney, "An Analysis of Four Methods for Measuring Pump Fluidborne Noise Generation Potential," *Proceedings of the National Conference on Fluid Power*, 31: 18–23, Oct. 1977.

20. BS 6335, *Methods for Determining Pressure Ripple Levels Generated in Hydraulic Fluid Power Systems and Components; Part 1. High Impedance Method for Pumps*, British Standards Institution, London, 1983.

21. *Hydraulic Fluid Power—High Impedance Method for Determining Pressure Ripple Levels of Pumps*, ISO Draft Proposal, document N221, submitted to Technical Committee 131, Sub Committee B, Working Group 1, International Standards Organization.

7
Pump Structureborne Noise

Pump structureborne noise, vibration, is important for the same reason that fluidborne noise is important. It causes other machine structural elements to vibrate and radiate airborne noise. It also often has 1000 times the energy of the pump's airborne noise. It is caused by all pump structural vibration modes, including those that do not produce sound.

Some machine structures are good at transmitting such energy and may even resonate and reinforce it. Transmission is by bending or torsional vibrations or by compression or shear waves that travel through solid members like sound waves travel through air. When their energy reaches a responsive member that is also a good sound radiator, high airborne noise levels are created. By converting only a fraction of 1% of the pump structureborne noise into sound, a member can radiate more airborne noise than the pump.

7.1 STRUCTUREBORNE NOISE MECHANICS

Newton's second law requires that when external periodic forces and moments act on a body, it vibrates. In the case of pumps, the principal outside influence is the small amount of shaft torque ripple. As we will see in a later chapter, there are also small forces generated by the hydraulic lines connected to the pump. These account for only a small fraction of the pump housing vibration that we refer to as structureborne noise.

Most of the structureborne noise is due to the mechanism shown schematically in Figure 7.1. The small mass m is a part of the pump acted on by internal pumping forces that displace it relative to the rest of the pump mass M. The swash plate in a variable displacement

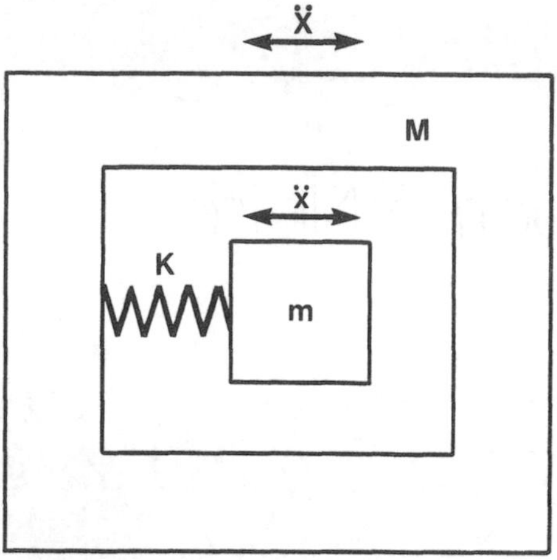

FIGURE 7.1 Principal structureborne noise mechanism.

inline piston pump is a good example. When acted on by piston forces, the yoke supporting it deflects so that the swash plate mass, as well as some of the yoke mass, is displaced relative to the rest of the pump. Newton's first law requires that the center of the total pump mass remain stationary unless acted on by an external force. The undeflected portion of the pump mass, then, must move in the opposite direction to compensate and keep the center of the total mass stationary.

Because the center of gravity remains stationary, the greater part of the pump mass moves a fraction of the distance moved by the deflected mass, relative to the center of gravity. This fraction is the ratio of the small to large masses. As a matter of fact, the velocities and accelerations of the two masses have the same relationship

$$X = \frac{xm}{M}$$

$$\dot{X} = \frac{\dot{x}m}{M}$$

$$\ddot{X} = \frac{\ddot{x}m}{M}$$

where

x, $\dot{x}$, and $\ddot{x}$ = absolute displacement, velocity, and acceleration of deflected masses

X, $\dot{X}$, and $\ddot{X}$ = absolute displacement, velocity, and acceleration of rest of pump mass

m = deflected mass, kg

M = balance of pump mass, kg

The relative motion within the pump is not limited to structural deflections. Functional motions of parts such as pistons and vanes cause structureborne noise. Their motions are large in comparison to those due to deflections, so they have a high potential for generating pump vibration. In the case of pistons, however, if their weights are equal, their inertias cancel like the vectors in Figure 3.8. If their weights are not equal, cancellation is not complete and there is structureborne noise in the direction of the shaft axis, at shaft frequency.

Vanes have a similar noise potential. They are also better matched, so they cancel more completely and seldom contribute to structureborne noise.

Any imbalance in rotating parts also causes structureborne noise. Most of these parts are fully machined, so pump rotating assemblies generally have little imbalance. Since imbalance produces forces at shaft frequency, which is always low, it rarely causes noise problems, even when it is abnormally high.

Imbalance does cause problems in pumps built for the Navy, however. Sonar-equipped ships have very stringent structureborne noise limits that sometimes can be met only by dynamically balancing the pump rotating group.

7.2 STRUCTUREBORNE NOISE SPECTRA

The generation and transmittal of structureborne noise are entirely different from those of fluidborne noise. However, their frequency spectra are very similar. This does not mean that for a given pump the spectra for the two forms of periodic energy are the same. Their similarity is a generalization which leads to referring to the typical fluidborne noise spectrum in Figure 6.6 as also being typical of structureborne noise spectra. Although structureborne noise mechanics provide some frequency weighting, this bias is less than that for airborne noise.

7.2.1 The Dynamic Factor

Structureborne noise is measured in terms of acceleration. It arises from sinusoidal forces that can be expressed in the form $P = p \sin \omega t$. As we saw in Section 5.1.1, this causes a vibration with a displacement

$$x_r = \frac{(DF)p \sin \omega t}{K} \quad \text{m}$$

where

DF = dynamic factor governed by the ratio of the forcing to natural frequencies

K = generalized stiffness factor, N/m

p = force amplitude

ω = force angular frequency, rad/sec

Differentiating provides the vibration velocity

$$\dot{x}_r = \frac{\omega (DF)p \cos \omega t}{K} \quad \text{m/sec}$$

Another differentiation provides the acceleration

$$\ddot{x}_r = \frac{-\omega^2 (DF)p \sin \omega t}{K} \quad \text{m/sec}^2$$

This acceleration is that of the deflected mass relative to the rest of the pump mass. The remainder mass has a countermotion, so the absolute deflection acceleration, measured from a stationary reference, is less than this amount.

As stated in Section 7.1, the absolute accelerations of the deflected and remainder portions of the pump are inversely proportional to their masses

$$\ddot{X}M = \ddot{x}m$$

The sum of these accelerations is equal to the initially calculated deflection acceleration $\ddot{x}_r$

$$\ddot{x}_r = \ddot{X} + \ddot{x} \quad \text{m/sec}^2$$

From this we find that the structureborne noise generated by the assumed sinusoidal force p is

$$\ddot{X} = \frac{\ddot{x}_r m}{M + m} \quad m/sec^2$$

$$= \frac{\omega^2 m(DF)p \sin \omega t}{(M + m)K}$$

The dynamic factor term DF reminds us that structureborne noise is affected by the relationship between the forcing and natural frequencies that governs all forced vibrations. The presence of a resonance has the same effect on pump structureborne noise as it has on airborne noise. There is one difference, however. Almost any resonance in a pump causes significant structureborne noise, while only those contributing to pump housing vibration modes that are good radiators significantly affect airborne noise.

Controls and other appurtenances attached to pumps are a common source of structureborne noise enhancements. When these are cantilevered from the otherwise "solid" pump structure, they often resonate. This seldom leads to airborne noise problems but causes difficulties in passing government structureborne noise tests. In cases where such tests are required, it is best to eliminate as many such structural appendages as possible.

7.2.2 Acceleration Factor

The expression for structureborne noise developed in the preceding section includes an ω^2 term. This indicates that the noise-generating mechanisms have a frequency bias equal to the square of frequency. Although this is greater than the linear velocity weighting for airborne noise, it is not augmented by a mechanism like radiation efficiency. The total frequency weighting is therefore less than for airborne noise. This accounts largely for the significant shaft frequency related peaks and the predominant pumping fundamental that are characteristic of structureborne noise spectra.

7.3 STRUCTUREBORNE NOISE MEASUREMENTS

Structureborne noise is the acceleration that a pump housing experiences if it is totally unconstrained. Although displacement, velocity, or acceleration describe its level, acceleration is used because it is easily measured. Expressed in decibels it is

$$L_{SBN} = 20 \log \frac{\ddot{X}}{\ddot{X}_0} \text{ dB}$$

where

$\ddot{X}$ = rms level of structureborne noise, m/sec^2

$\ddot{X}_0$ = reference acceleration level, 10 μm/sec^2

Structureborne noise measurements are not commonly made, although there are a number of good reasons for making them. The foremost reason is to compare the outputs of competing pumps when minimum noise is sought. A corollary to this is to demonstrate the compliance with a government purchase specification. Structureborne noise is more sensitive than airborne noise to pump fabrication variations, so it is useful as a final production check and for monitoring pump condition in service. With carefully designed testing, it is also a powerful diagnostic tool.

There are no standards for measuring pump structureborne noise, nor are there any indications of efforts to develop one. Probably this is because it is recognized as a difficult task. Some of the people who developed the airborne standard and are involved in developing a fluidborne standard, however, indicate that a structureborne standard is just waiting its turn.

There is one structureborne noise measurement standard, MIL-STD 740B (1), that provides some guidance for making such measurements. Unfortunately, it is for making measurements on entire machines. Although it works fairly well for hydraulic power supplies, consisting of a pump, drive motor, and ancillary equipment, it is not directly applicable to pumps alone.

There are several reasons why pump structureborne noise cannot be measured accurately. Understanding these reasons is essential to making practical measurements that are adequate for meeting ad hoc objectives.

7.3.1 Test Setup

The structureborne noise analysis given earlier in this chapter assumes that part of the pump is deflected a certain amount relative to the remainder of the pump. This would be the case if the pump consisted of two rigid bodies connected by a resilient member, as shown in Figure 7.1. Unfortunately, the pump housing whose motion we wish to measure is not rigid. Its distortions have about the same magnitude as its structureborne noise. Because of this, measurements made at different points on the housing produce different statistics.

Common practice is to mount the object whose vibration is being measured on a rigid steel, cast iron, or concrete block. This, in turn, is resiliently mounted so that the assembly consisting of the pump and its supporting rigid body have a natural frequency well below the lowest measurement frequency. Motion of the assembly as measured on the rigid block is then proportional to the pump's inherent structureborne noise and the assembly's total mass. This practice is referred to as *seismic mounting.*

This practice has one obvious drawback, especially in the case of pumps. The rigid body has far more mass than the pump and therefore has a great influence on the results. It is difficult to correct the data for this added mass. In fact, government purchase specifications generally ignore this problem and require no correction.

Pumps often are flange mounted on angle brackets that support them on the rigid body. Even massive brackets vibrate, and this induces errors. Sometimes, to minimize this error, the measurements are made at the pump mounting flange. This appears to add to the arbitrary nature of the measurements.

Generally, pumps are driven through a flexible coupling to ameliorate misalignment. This is not adequate when the pump is resiliently mounted. A single coupling provides too much restraint to seismic mass movement. For this reason the pump must be driven through two flexible couplings separated by a section of shaft. The coupling spacing is not critical, but more is better.

Even with two couplings it is necessary to align the motor and pump shafts carefully. Misalignment causes nonuniform shaft rotation, and this increases the structureborne noise. Parallel misalignment increases the second shaft harmonic and angular misalignment increases the fourth. Alignment should be done with all lines attached to the pump and filled with oil so that the resilient mounts are fully deflected.

7.3.2 Instrumentation

Theoretically, pumps can have periodic forces acting along three mutually perpendicular coordinate axes and three periodic moments about the same axes. So a seismically mounted pump can have six different oscillations: three linear and three rotational. These are shown in Figure 7.2. No pump type has all of these motions, but a universal measurement system must be able to gauge them all.

An accelerometer measures total acceleration in the direction of its measuring axis. For a body with all six motions, this total usually includes accelerations due to rocking about the two axes perpendicular to the measuring axis as well as that of the linear motion in the direction of the measuring axis. For many purposes this does not cause problems.

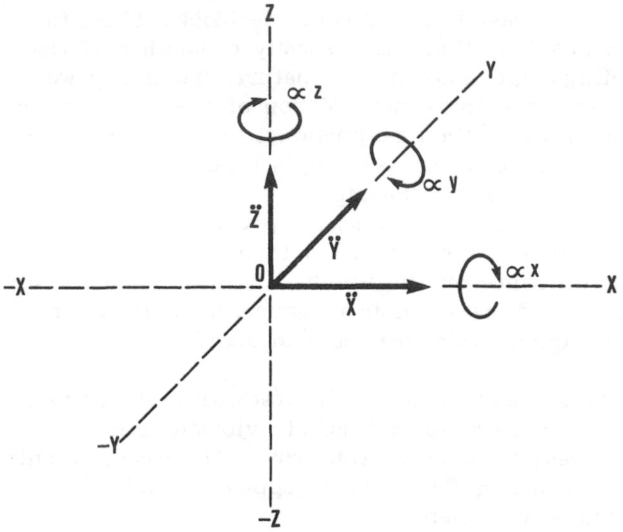

FIGURE 7.2 Six possible components of pump structureborne noise.

Triaxial arrays of accelerometers are frequently used for structureborne noise measurements. These have three transducers oriented to measure accelerations along three perpendicular coordinate axes. Generally, they are oriented so that one axis is parallel to the shaft and one is vertical. Often pairs of these arrays, located at diametrically opposite mounting points, are used.

Some arrays are a single transducer with the three accelerometers built into them. More often, however, they are made by attaching three accelerometers to a steel cube with studs. The cubes are either welded or brazed to the seismic mass or pump support bracket. Under some circumstances, the blocks are epoxied in place.

Special insulated studs must be used to attach transducers to the blocks. The shielding of the wires connecting the accelerometers to the instrumentation is grounded to both the transducers and instrument. If insulated studs are not used, all these shields are connected together at the blocks as well as at the instrument. They therefore form electrical loops. Stray magnetic fields cutting these loops generate electrical currents in the shielding. These currents, in turn, induce spurious signals in the measuring circuits.

Although insulated studs eliminate this electrical error, they can cause mechanical errors. They are relatively fragile and must be handled very carefully. They must also be checked frequently to see if they are causing erroneous readings. This is usually done by replacing the stud with a fresh one, known to be good, or by switching studs around to see if noise-level readings change.

Insulated studs are not needed if measurements are made with a single accelerometer that is moved to the various positions. A magnet is sometimes used to hold the transducer and facilitate moving it about. Magnetic mounting on relatively flat, smooth surfaces is generally satisfactory up to 2000 Hz and is therefore satisfactory for most noise evaluations. Epoxy, Eastman 910, and dental cements are also satisfactory over this frequency range if care is taken to keep their thickness at a minimum. These are only rarely used, however, because of the curing delays that they incur.

7.3.3 Test Techniques

Selecting a method of measuring structureborne noise depends on the purpose. When comparing pumps or pump modifications, using the same test set up for all units, the highest total level from one of the accelerometers may be adequate. A refinement of this is to take the vector sum of the three orthogonal accelerations at an array. This vector sum is found by adding the total levels from the three transducers, expressed in decibels, in the same way that sound power levels are added.

If certain frequencies are critical, levels of these frequenices or the third-octave bands that include them should be measured instead of the total levels. This will greatly increase the test discrimination.

Tests to identify sources of structureborne noise require considerable ingenuity and cannot be standardized. Although difficult to make, these tests are based on a rather simple strategy.

Signal components from an accelerometer mounted on a vibrating body having complex motion are shown in Figure 7.3. Point O in this figure is the center of gravity of the test assembly, which includes the pump, mounting bracket, and rigid body. It is common practice to position the resilient members supporting the assembly symetrically around this point so that it is also the center of rotation. Most of the mass is in the rigid body, and if it is a simple shape, the location of its center of gravity is accurately known. The other components are much lighter, so their centers of gravity do not have to be known very accurately to pinpoint the assembly center of gravity.

As seen in Figure 7.3, the measured acceleration is the sum of the linear acceleration of the test assembly in the direction of the accelerometer's axis, and two accelerations in the same direction that arise from the assembly's rotational oscillations

$$A = \ddot{X} + Z_1\alpha_Y + Y_1\alpha_Z \quad \text{m/sec}^2$$

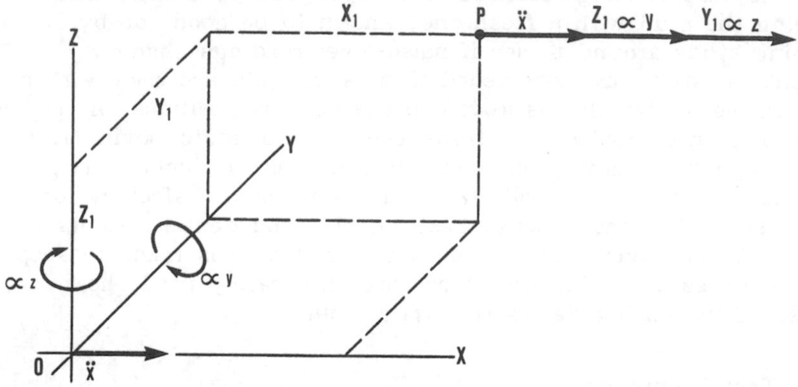

FIGURE 7.3 An accelerometer generally measures the effects of three accelerations.

where

$\ddot{X}$ = linear acceleration of point O, m/sec^2

Y_1 = Y coordinate of measuring point, m

Z_1 = Z coordinate of measuring point, m

α_Y = angular acceleration about the YY axis, rad/sec^2

α_Z = angular acceleration about the ZZ axis, rad/sec^2

If the measuring point is located where one of its coordinates is zero, such as in the XY plane in this case, the contribution of one of the rotations is zero.

It generally is not possible to locate the accelerometer where the effects of both rotations are eliminated. The effect of the second is eliminated, however, by using another accelerometer, located as shown in Figure 7.4. Here the direction of the rotation acceleration vector is reversed. At one measuring point the rotation and linear motions add, and at the other they oppose each other. Connecting the outputs of the transducers in series, so that their signals add, the rotational components cancel and the signal is twice the linear acceleration. By reversing the polarity of one of the accelerometers, the linear motions cancel and the output of the transducers is twice

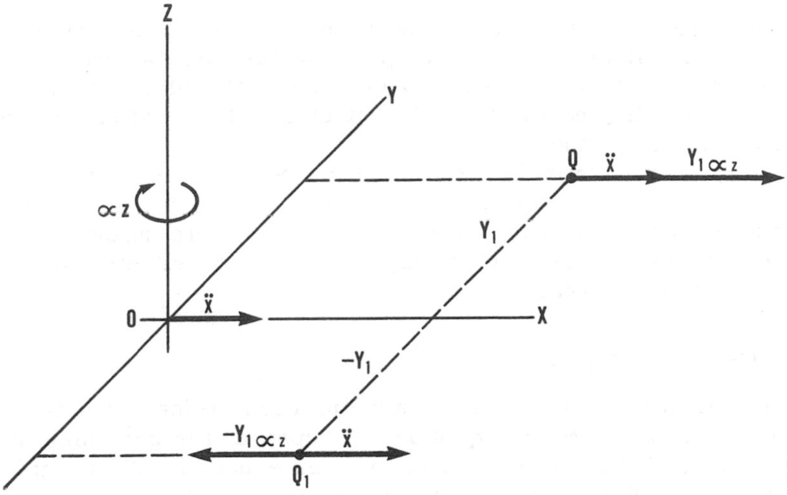

FIGURE 7.4 Two accelerometers can be used to separate acceleration components.

the rotation acceleration. The rotational acceleration is then found by dividing by the distance between the accelerometers.

A pair of measuring points only serve for one measuring direction. Three sets of points are needed to measure all six motions of the seismic mass assembly. Generally, only the strongest motions are of interest, and measurements at one or two pairs of points are all that is required.

Once all the motions are defined, computions can be made to eliminate the effects of the rigid body and mounting plate. Generally, this is not useful.

7.4 DESIGN CONTROL

Pump structureborne noise is caused by the same periodic pumping forces and moments that produce airborne noise. For this reason it is reduced in many of the same ways that are discussed in Chapter 5 for controlling pump airborne noise. Because these two types of noise have different sensitivities to frequency, however, design modifications do not affect them to the same degree.

7.4.1 Port Timing

Port timing changes that reduce airborne noise also reduce structureborne noise. However, since the pumping fundamental frequency of this noise is highest and such changes have little effect on the fundamental, porting modification will not change the total structureborne noise level by much.

Generally, our interest in structureborne noise is due to its ability to generate airborne noise. In such cases, reductions in the mid-frequency harmonics critical to airborne noise are more important than the total level. Porting changes, of course, are effective in reducing these harmonics.

7.4.2 Effect of Stiffness

Structural stiffness is important in structureborne noise. This is the factor K given in the motion equation developed at the beginning of this chapter. Although its prominence in this equation indicates its importance, it is also useful to consider it in terms of noise energy.

When a periodic force acts on a structure, it stores potential energy within the structure that is equal to the integral of the product of the instantaneous force times the resultant deflection. As shown in Figure 7.5, this integral is the area under the deflection lines. Because of linearity, this area is equal to one-half the product of the maximum force and deflection. When the maximum force is reached, motion stops, so there is no kinetic energy at this instant and the area is equal to the system's total vibratory energy.

From Figure 7.5 we also see how increasing stiffness decreases the area under the graph lines. This often leads to significant noise reductions without increasing pump cost. In bending, for example, stiffness increases with the cube of thickness so that only a small increase in material has a big effect. Utilizing this approach to reducing noise, however, requires identifying its source and the stiffnesses involved.

This discussion on the effect of stiffness assumes that the forcing frequency is well below the structure's resonant frequency. Near and above the resonant frequency the deflection and force are out of phase, so their product is no longer twice the potential energy of the vibrating system, as shown in Figure 7.5.

Above resonance, motion is proportional to the deflected mass. Although increasing this mass decreases its amplitude, it increases the mass driving the whole pump. The net effect is therefore zero.

Increasing stiffness of a structure increases its natural frequency. When a system is vibrating above its resonant frequency, the change must be great enough to shift this frequency well above the forcing frequency; otherwise, it will increase the dynamic factor and the vibration.

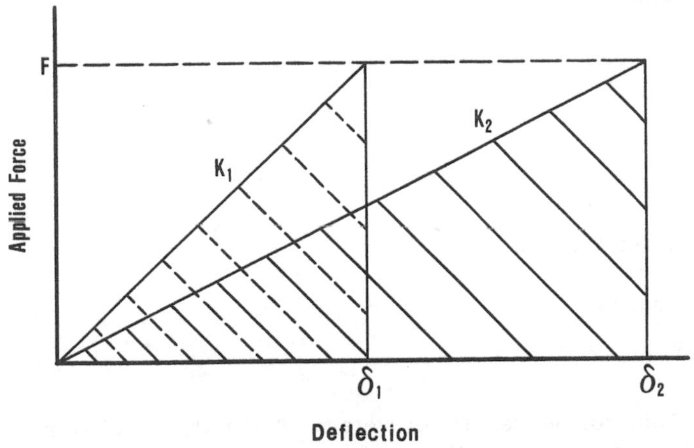

FIGURE 7.5 Vibratory energy decreases as stiffness increases.

Fluid columns such as those in pump displacement control cylinders are common sources of flexibility. Since compressibility is a volume phenomenon, it is often assumed that decreasing the fluid volume will stiffen it. This is not always true.

The spring rate or stiffness of an oil column is

$$K = \frac{F}{\delta} \quad \text{N/m}$$

where

F = force applied to piston, N

δ = piston motion caused by force, m

This deflection is equal to

$$\delta = \frac{\Delta V}{A} \quad \text{m}$$

where

ΔV = volume change due to compressibility, m^3

A = piston area, m^2

The change in volume is

$$\Delta V = \frac{VF}{AB} \quad m^3$$

where

B = fluid bulk modulus, Pa

V = enclosed fluid volume, m^3

The volume is equal to

$$V = AL \quad m^3$$

where L is the fluid column length in meters. When these relationships are combined, we see that the stiffness is

$$K = \frac{AB}{L} \quad N/m$$

From this we see that stiffness is increased by decreasing the length, which reduces the enclosed volume. However, it is also increased by increasing the piston area, and this increases cylinder volume.

REFERENCE

1. MIL-STD-740B (SHIPS), *Airborne and Structureborne Noise Measurements and Acceptance Criteria of Shipboard Equipment*, January 13, 1965.

8
Valve Noise

Valves generally do not produce noise levels as high as those of pumps. However, in some cases, they are the predominant noise source in a given machine. Even where their noise levels do not exceed the pumps, this noise can often be distinguished from the pump noise and adds to the perceived noisiness of a machine.

Valve noise grew in importance as more mobile machines were equipped with operators' cabs. These cabs contain the noise energy of operator-controlled valves instead of letting it escape. In some cases the cab walls also became valve noise radiators. Adding cabs to these machines, then, raised operator ambient noise levels to where they became an annoyance and in some cases, endangered operator hearing. Valve noises that were acceptable in the past are now the object of concern.

Three mechanisms cause valve noise: cavitation, instability, and impact. In this chapter we also consider hydraulic shock initiated by valves, even though this is not strictly valve noise. Although these mechanisms are not of equal importance, they are all the source of problems so a familiarity with them is important in hydraulic noise control.

8.1 VALVE TYPES

This is not an attempt to catalog all valve types in use. Only the few needed to provide a common background and to illustrate details of noise mechanisms are discussed here.

Figure 8.1 shows three common types of pressure control valves. In the first, control is provided by a spring-loaded poppet. Cones and balls are usually used for poppets, but a variety of other

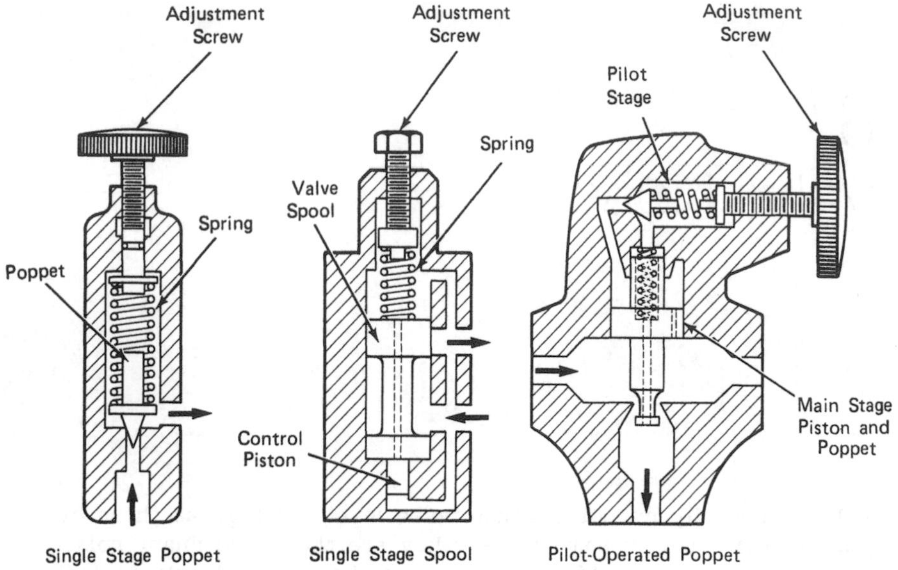

FIGURE 8.1 Pressure relief valves.

shapes are used as well. When the pressure acting on this control element exceeds the *cracking pressure*, for which the spring preload was set, it is unseated and flow through the orifice keeps the pressure from rising further.

As soon as the poppet lifts off its seat, pressure acts over a larger surface area, so the poppet opens wider than necessary to maintain the cracking pressure. The controlled pressure is therefore lower than the cracking pressure. Further, the spring is relatively stiff, so the controlled pressure increases as the poppet rises to accommodate higher flows. This type of valve also tends to be unstable. Such valves are most often used as relief valves where these weaknesses do not cause problems.

The second valve utilizes a spool element, which generally can handle larger flows than a simple spring-loaded poppet. Pressure forces acting on the spool are always balanced. Only inlet pressure acts on the two inboard faces and a hole through its axis equalizes the pressure on the ends. The spool is held in the closed position by a soft spring whose preload is adjusted to set pressure.

The spool is shifted by inlet pressure acting on the small piston at its lower end. When the inlet pressure exceeds the set pressure, the piston moves the spool to open a flow path to the discharge port. Once opened, any pressure increase causes the piston to move the spool to increase the opening, and this compresses the spring further. This action causes the controlled pressure to increase with flow, although this increase generally is less than that of simple poppet valves.

These valves have more damping than poppets and so are generally more stable. They still tend to be unstable, however. Like simple poppet valves, they are most frequently used for pressure relief rather than pressure regulation.

The third valve is perhaps the most common, although it is more expensive than the other two. It is referred to as a two-stage or pilot-operated valve. The first stage of this valve is a spring-loaded poppet. The second stage is either a spool or a poppet, driven by a piston. The one shown in Figure 8.1 consists of a one-piece piston and poppet, held in the closed position by a relatively soft spring. A small hole in the piston land balances the pressure on the two piston faces. This hole also provides the fluid path to the first stage.

When pressure at the valve inlet exceeds the set pressure, the first-stage poppet is unseated. Flow to the valve produces a pressure drop as it passes through the small hole in the piston. This reduces the pressure on top of the piston so the piston moves upward, against the spring, opening the main orifice and letting fluid flow from the inlet to the discharge port. Fluid flowing through the first stage also reaches the discharge port through an axial hole through the piston and poppet.

If the pressure continues to rise, flow through the pilot stage increases, increasing the pressure drop in the hole through the piston. This drops the pressure on the upper piston surface, causing the piston to move until the spring is compressed enough to restore equilibrium. This opens the main orifice wider to accommodate additional fluid flow. Since the spring is relatively soft, this action maintains pressure at the valve inlet within a narrow range of the set pressure.

The shape of the lower end of the poppet was developed to compensate for the flow forces reacting against the poppet. Because of it, flow has a minimum effect on the controlled pressure.

The popularity of pilot-operated valves is not due only to their better pressure control characteristics. They are also generally much more stable because they have greater damping and are somewhat slower acting.

Spool and poppet metering elements can also be actuated by discharge port pressure. When they are, they become reducing valves, which are commonly used when one branch of a hydraulic system must operate at a pressure lower than the rest of the system. Similarly, they can be actuated by the pressure differential generated by the controlled flow passing through an orifice. They are then flow control valves that maintain flow at a constant, set level. Most pressure-reducing and flow control valves operate with relatively small pressure drops in their main stages, so usually do not generate high noise levels.

Another type of valve that is of interest from a noise standpoint is the directional valve. These are generally of the spool type. However, the use of poppet types appears to be gaining in popularity because poppets provide a better seal. This is especially important with high-water-content fluids, which because of their very low viscosities, leak readily. It is also important in machines like cranes, where leakage allows the load to fall.

Directional valves are used to direct flow to one side of an actuator, such as a hydraulic cylinder or hydraulic motor, while providing a drain to tank for the other side. Switching the valve reverses the connections and the direction of flow through the actuator, thereby changing the direction of the cylinder travel or motor rotation. These valves are actuated manually, by solenoid, and hydraulically. Figure 8.2 shows a manually operated unit which is useful in some types of machinery where a degree of manual flow or speed control is also desired.

A spring holds the spool of this valve in its farthest-left position. Pressurized fluid entering port P is directed to actuator port A while the other actuator port, B, is connected to the tank port T to drain fluid back to the system reservoir. When the handle is

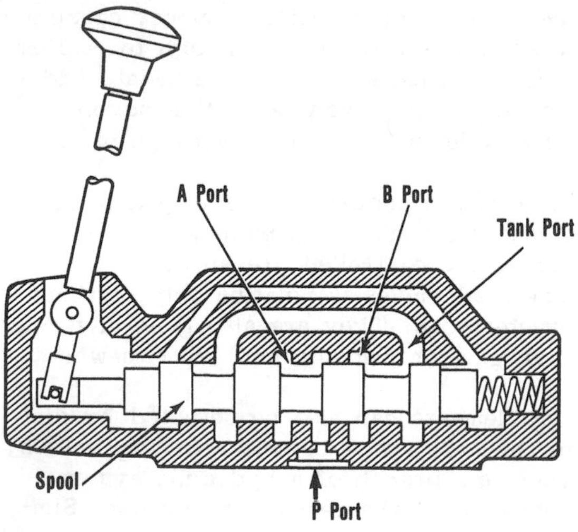

FIGURE 8.2 Spring-offset manual directional control valve.

moved to the left as far as it can go, the connections are reversed; B is connected to the pressure source and A is connected to the tank port. In these two positions the pressure drop through the valve is as low as the manufacturer could make it, so there is little potential for generating noise.

Directional valves do not always operate in their fully opened positions. Quite commonly with hand-operated valves, the operator uses it to control speed by opening the valve only partially so that the flow is throttled. The pressure drop across a valve that is barely opened or feathered is high, so most directional valve noise problems occur during this type of operation.

Perhaps it should be pointed out that there are two basic types of directional valves. The one that was just described is termed a four-way valve. There are also three-way valves. These have a similar construction and are used with actuators having a single port, such as a single-acting cylinder. Their function is to switch the pressure port or the tank port to the actuator, as required.

8.2 CAVITATION

Cavitation is by far the leading noise-generating mechanism in valves. It is quite common in pressure-regulating and pressure relief valves. When these valves cavitate for a long time, they may be more of an erosion problem than a noise problem. When erosion occurs, it not only reduces valve life but pollutes the fluid with metal particles, which cause pump damage.

8.2.1 Cavitation Mechanics

Valves control fluid flow by constricting the flow path, as shown in Figure 8.3. This causes the fluid to speed up in passing through the constriction. Since friction losses in turbulent flow are proportional to the square of velocity, the needed energy loss is then achieved in a relatively short distance.

In accelerating, some of the fluid's potential energy is converted to kinetic energy, in accordance with Bernoulli's theorem. The pressure in the restriction is therefore reduced as shown in Figure 8.3. The jet leaving the constriction interacts with the static fluid in the valve discharge to form eddies. Because of centrifugal action, the pressure in the center of these little whirlpools is lower than even the jet pressure. Further downstream, where the jet is able to dissipate and resume a slower flow, some of the lost pressure is recovered.

The jet persists for some distance down the valve discharge line. If the jet faces straight down the discharge line, it acts as an

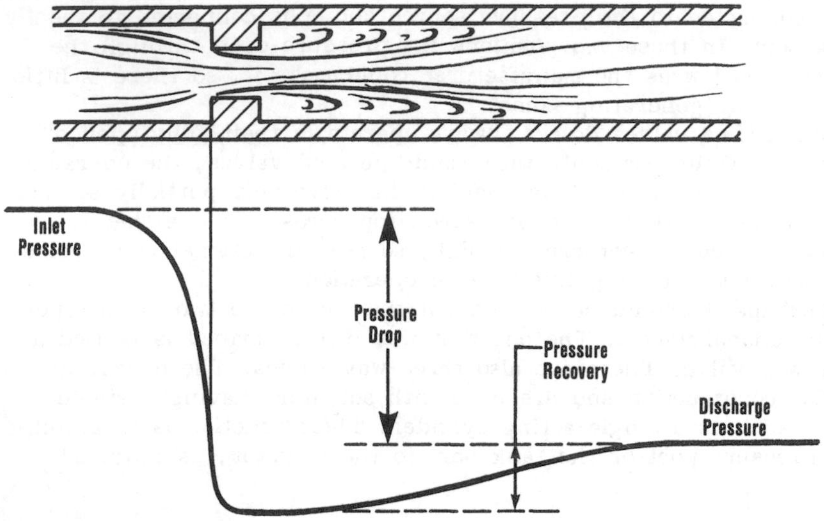

FIGURE 8.3 Orifice pressures.

ejector, at very high flows, and can create a void in the line by drawing out the static fluid (1). It is interesting that the researchers who reported this phenomenon also noted that the valve noise was emitted downstream where the jet dissipated and not at the valve.

8.2.2 Critical Pressures

Fluids cavitate whenever their pressures fall below critical levels. Each fluid has two different cavitation initiation pressures, one where outgassing starts and the other where the fluid begins to vaporize.

Outgassing is the release of air or other gas that is dissolved in a fluid. Figure 8.4 shows how much air can be held in solution by petroleum- and water-based fluids when these fluids are held at various pressures for a long period of time at room temperature. The soluability of the gas given in this chart is the volume that the gas would have at 32°F and atmospheric pressure, divided by the volume of the fluid.

This dissolved air does not affect the fluid's other physical properties, such as viscosity or bulk modulus. However, whenever the fluid pressure drops below the *saturation pressure* corresponding to the amount of dissolved air, cavitation or outgassing occurs. The

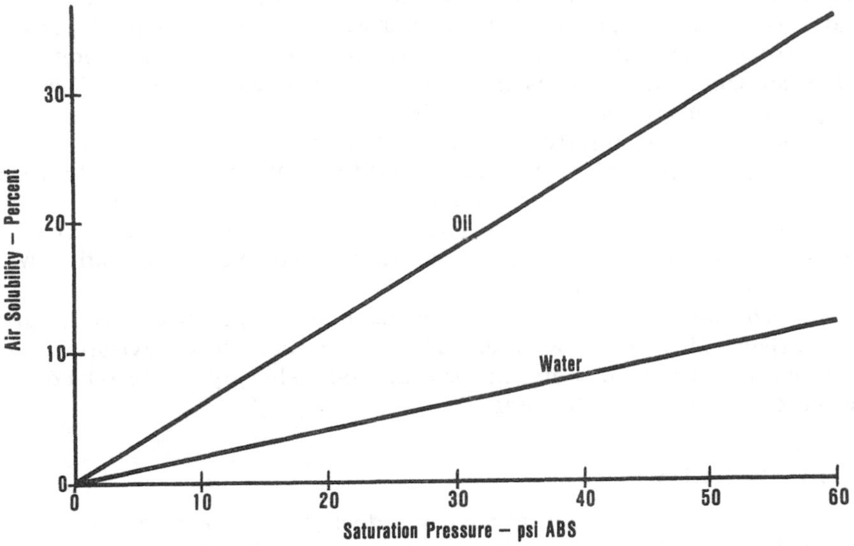

FIGURE 8.4 Air solubility in oil and water.

further the fluid pressure goes below this critical pressure, the more violent the bubbling. Given enough time, this action stops when the dissolved air drops to the level commensurate with the lower pressure.

The saturation pressure at a valve depends on details of the rest of the hydraulic circuit. It tends to be close to the reservoir pressure because the fluid spends much of its time in contact with air in this location. Whereas this is atmospheric pressure in most machines, it can be much higher in circuits with pressurized reservoirs. Where the supply pump injests entrained air, either drawn from the reservoir or leaked through suction-line joints, this air dissolves in the fluid when it is pressurized. In such cases the critical pressure of the fluid reaching the valve is above reservoir pressure.

In contrast, some of the cavitation air produced by a valve may escape in the reservoir and not be replaced by absorption before the oil is recirculated. In such cases the critical pressure at the valve is close to atmospheric at startup but drops to a lower level when the machine operates for a while.

The *vapor pressure* is the pressure below which a fluid will begin to vaporize. It is generally below the fluid's saturation pressure. The vapor pressure for oils is below 0.01 psi absolute. Water has a vapor pressure of about 2 psi absolute at a temperature of 125°F, and this rises to 8 psi absolute at 180°F. The lower the fluid pressure drops below the vapor pressure, the greater the cavitation violence. Once cavitation is initiated, it continues as long as the temperature and pressure remain the same.

When a valve is lightly cavitating, it is probably the result of outgassing. If the valve operating conditions are changed to depress the orifice pressure further, vapor cavitation may also occur. Since vapor bubbles have much more energy than air bubbles, they produce greater noise and erosion. For that reason vapor cavitation will overshadow gas cavitation.

Oils containing only a little water have two vapor pressures, with the higher being that for water. For this reason, mild cavitation normally occurring in a circuit may increase sharply if the oil becomes contaminated with water.

8.2.3 Cavitation Noise

As suggested earlier, cavitation itself does not cause noise. It is the collapse of the cavities that causes noise. With valves this occurs when the jet dissipates into a more normal flow and pressure recovery takes place. Often this happens in the discharge line, outside the valve.

Bubble collapse releases a surprising amount of energy. When it occurs at a solid surface, it is capable of causing surface fatigue failures, pitting, in all but the hardest materials. The energy also causes structural vibration that can end up as a loud noise. In addition, the reaction with the rest of the fluid results in high levels of fluidborne noise.

When cavity collapse occurs within the fluid, away from solid surfaces, it produces only fluidborne noise. No erosion occurs. The fluidborne noise, of course, transmits noise energy to structural elements so that air and structureborne noise are by-products, but they have much lower levels than when collapse occurs at a surface.

Cavitation-generated fluidborne noise is confined to the downstream side of the valve orifice (1). While the impedance of the orifice is sufficiently different from that of the downstream passage, to reflect much of this noise downstream, it is believed that the cavitation bubbles do most of the reflecting.

Cavitation noise is random, like the bubble collapse that causes it. Fluidborne noise exists at all frequencies but is strongest in the range 4 to 8 kHz (2). Because of its strong high frequencies, it is efficiently radiated as airborne noise. It is generally described as a hissing sound.

The best data on valve noise are found in Ref. 1. Some of these data are also given in Ref. 2. Figure 8.5, from these papers, shows

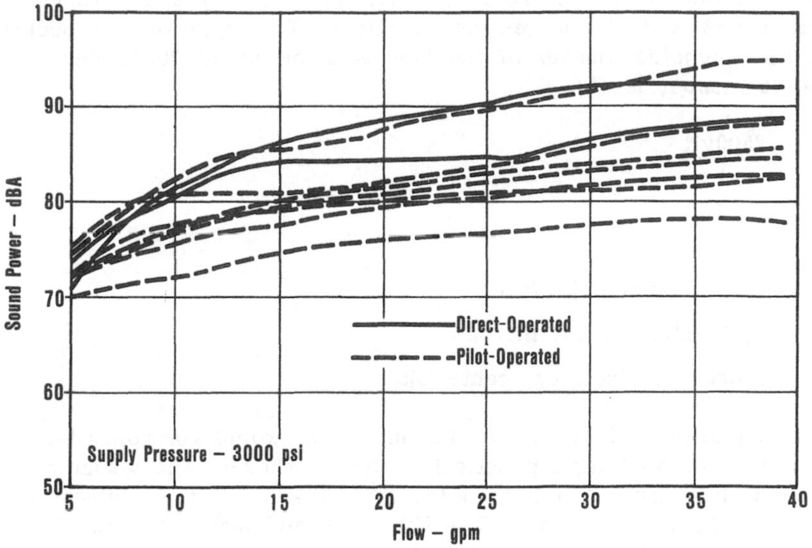

FIGURE 8.5 Noise levels of nine relief valves. (From Ref. 1.) Reprinted by permission of the Council of the Institution of Mechanical Engineers.

the sound power levels for nine relief valves and their discharge lines. These valves were operating with a supply pressure of 3000 psi and atmospheric back pressure. From these data we see that, in general, pilot-operated valves are quieter than single-stage valves and that there is 10 to 15 dB difference between the loudest and the quietest valves.

These data also show that except for low flows, sound power levels increase about 3 dB for each doubling of flow, indicating that the sound power is proportional to flow.

There are few data showing the effect of supply pressure on valve noise. Cross-plotting some data in Ref. 3 suggests that sound power increases with the square of pressure. This includes the noise from both the valve and its discharge line where bubble collapse occurs. Earlier data which did not include the line noise show that the noise radiating from valves is almost independent of supply pressure (4).

8.2.4 Cavitation Control

A good way to avoid cavitation is to design the throttling device so that it has laminar flow. The advantage of this type of flow is that it uses viscous friction to achieve fairly good energy loss with low velocities and it does not generate vortices. This type of flow occurs when the *Reynolds number* of the flow is below about 2000. The Reynolds number is (5)

$$N = \frac{2500vR}{\nu}$$

where

v = fluid velocity, in./sec

R = hydraulic radius, in.

ν = kinematic viscosity, centistokes

This equation is in terms of the units that would commonly be encountered in working a practical problem. Whereas the velocity and the radius are in English units, centistokes, used for viscosity, is metric. The constant reconciles the units and makes the resulting number dimensionless, as it should be.

The hydraulic radius, in this equation, depends on the shape of the flow cross section. It is defined as

$$R = \frac{\text{cross-sectional area, in.}^2}{\text{perimeter, in.}} \quad \text{in.}$$

In some laminar flow schemes, the fluid is divided into many small flows, whereas in others it flows through a film. Where circular passages are used, R = diameter/4, and for films that are very thin in comparison to their width, R = thickness/2.

From inspection of the Reynolds number equation it is seen that the allowable velocity is inversely proportional to either film thickness or path diameter. In the case of a film whose width is constant, reducing the thickness, the cross-sectional area is exactly offset by the higher allowable velocity. So the allowable flow through films is independent of the film thickness.

In the case of passages that are round or whose cross-sectional dimensions remain proportional, their areas decrease as the square of their size while their allowable velocity changes only linearly. It would seem desirable to keep the passages as large as possible to keep from having to provide large numbers to handle a given flow. However, as velocities are reduced, the passage length must also be increased, to achieve a given pressure drop. It can be shown that pressure drop for laminar flow, in the context of this discussion, is

$$\Delta p = k \frac{lv^3}{N_R} \text{ psi}$$

where

k = a constant

l = path length, in.

v = average flow velocity, in./sec

N_R = Reynolds number

From this it can be seen that the real advantage is in reducing passage size and increasing velocity, because this rapidly reduces the length needed for a given pressure drop.

Porous materials are sometimes used to achieve laminar flow. Both compacted stainless steel wool and sintered powdered metal have been used. These not only provide small pore sizes, they also offer paths with many turnings which are good for energy loss. However, they erode easily and shed debris if velocities are too high or if some cavitation occurs.

A difficulty with using this material is in making the valve adjustable. Figure 8.6 shows the general idea of an adjustment scheme. With this configuration, either the porous plug or the outer port sleeve can be moved back and forth to change the flow path length.

A similar throttling mechanism utilizes flow paths etched in the faces of washers. Multiple paths are provided by stacking many washers together to form a porous sleeve. The exact details of this valve are not recalled accurately, but the general arrangement was somewhat like that shown in Figure 8.7.

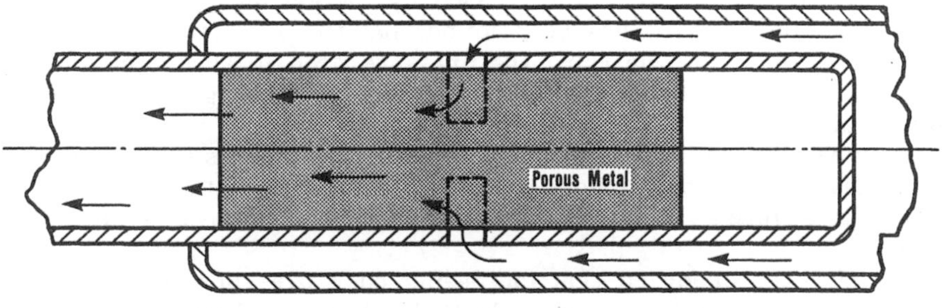

FIGURE 8.6 Porous metal valve concept.

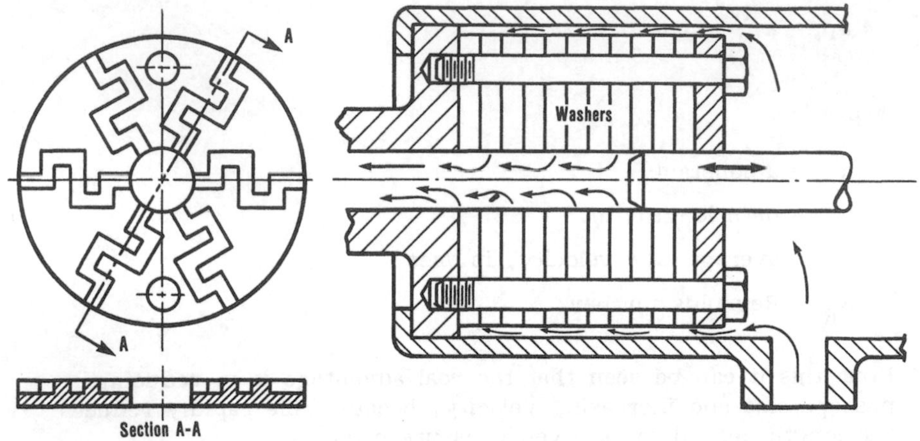

FIGURE 8.7 Etched washer valve concept.

Figure 8.8 shows a valve built at the National Engineering Laboratory (NEL) that utilizes a film flow to dissipate energy (6). The author used similar hand-operated valves in the laboratory for providing the load in pump noise evaluation testing. Although they were quiet, it was felt that they needed considerable mechanical refinement before they could be considered for commercial use.

NEL also developed valves that have only one or two flow passages for flows up to 300 gal/min (7). As shown earlier, if a path has a large cross-sectional area, a long path is necessary to get the needed pressure drop. They provided this by making the flow passage in the form of a helical thread. An ingenious system of intersecting holes makes the length of the passages adjustable with only a moderate amount of motion.

Since cavitation is caused by low pressure, it seems reasonable to expect that it could be suppressed by increasing a valve's back pressure. However, it has been found that the absolute back pressure must be about one-third the pressure drop to eliminate cavitation in conventional valves. This approach, then, is generally not practical.

Further, lower, more practical levels of back pressure do not reduce cavitation. They have a curious effect on cavitation noise, as shown in Figure 8.9. Increasing back pressure increases sound power.

It has been observed that increasing back pressure shifts the bubble collapse zone upstream. From this it is concluded that the sound increase is due to having more bubbles collapse near the solid valve surfaces. The maximum occurs when the collapse zone

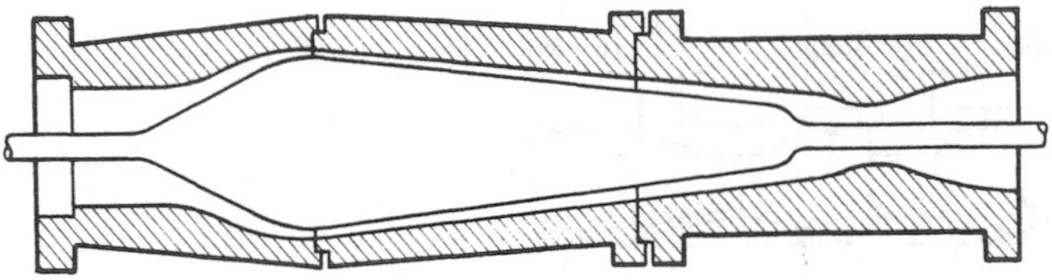

FIGURE 8.8 NEL cone-type laminar flow valve. (From Ref. 6.) Reprinted by permission of the Council of the Institution of Mechanical Engineers.

reaches the valve. The fact that this maximum noise occurs at back pressures that increase with flow appears to support this theory. Noise reductions occurring when back pressure is increased above the maximum noise pressure are probably due to reduced cavitation.

Some noise reduction is achieved by using a series of pressure drops, with the downstream ones providing back pressure to suppress cavitation in their upstream counterparts. A pilot-operated valve was built with two poppet valves in series with the control, dividing the pressure drop equally between the two stages. This valve produced 9 dB less noise than that of a comparable valve of

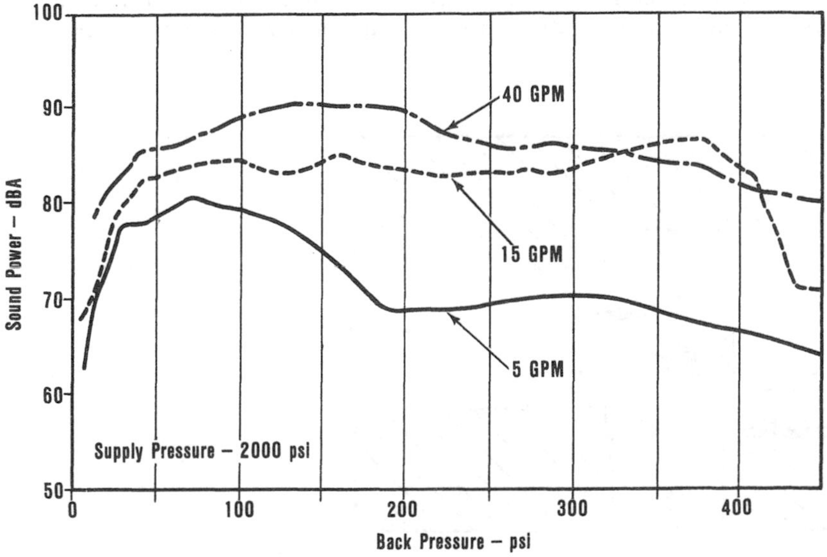

FIGURE 8.9 Effect of back pressure on relief valve noise. (From Ref. 1.) Reprinted by permission of the Council of the Institution of Mechanical Engineers.

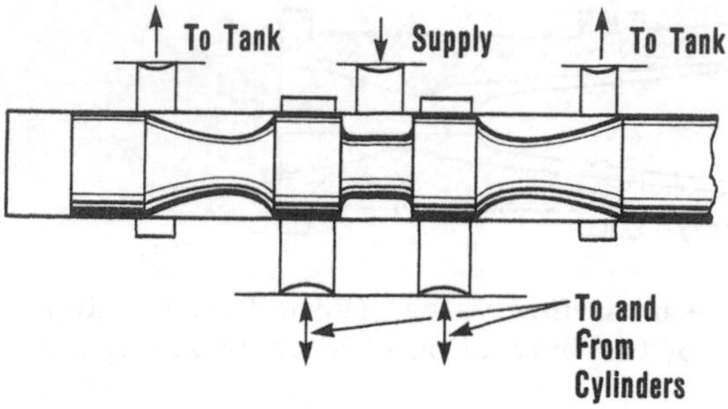

FIGURE 8.10 Valve spool contoured for low noise. (From Ref. 6.) Reprinted by permission of the Council of the Institution of Mechanical Engineers.

conventional design. The fact that this scheme was so successful reinforces the perception that total valve cavitation noise levels are a function of at least the square of the pressure drop.

Cutting valve cavitation noise does not require eliminating cavitation. Reductions, perhaps up to 5 dB, have been made by recontouring the throttling device and valve passages. The objective of these modifications is to move cavitation collapse farther from solid surfaces and to smooth discharge flow to reduce vortex formation. The author has seen some promising data where the lower end of the poppet of the pilot-operated valve in Figure 8.1 was recontoured.

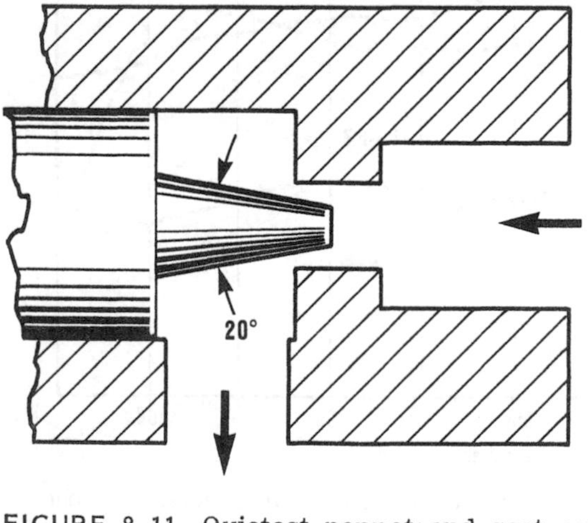

FIGURE 8.11 Quietest poppet and seat combination. (From Ref. 8.)

It is believed that noise reductions achieved by contouring the spool valve shown in Figure 8.10 is due to this factor.

Throttling element modification can produce larger noise reductions. Tests made with four different cone-shaped poppets in combinations with three seat shapes found that these combinations produced a noise level range of 13 dB (8). Figure 8.11 shows the combination with the lowest noise level. There was a definite increase in noise as the poppet angle was increased. The quietest seat was the straight-sided orifice shown. Convergent and divergent orifices had little effect on noise levels for the low-angle poppets, but caused 8-dB increases with the 100° poppet.

It was found that part of the reason for the high noise levels with the large-angle poppets was due to the flow restriction that they caused. Reducing the poppet diameter to provide a more generous downstream passage reduced the noise by 4 dB.

The most important finding was that cavitation and noise could be greatly reduced by having the jet strike a solid surface soon after leaving the orifice. This appears to be effective because it initiates an early pressure recovery. Since the time available for bubble formation is shortened, fewer are formed. Applying this concept to an already quiet poppet configuration reduced the noise level nearly 5 dB, which made it quieter than any of the comparable commercial valves that were tested.

This noise reduction scheme can be overdone. It was found that if the impingement surface is too close to the orifice, some of the pressure drop occurs between it and the orifice face. Cavitation then occurs when the fluid leaves this area and little suppression is achieved.

8.3 VALVE OSCILLATION

This noise is a single-frequency or pure tone sound, generally described as a squeal or whistle. Fortunately, it is infrequently encountered, although it has been the subject of much research (9–11). A lot is known about its mechanics, but it is difficult to determine analytically what must be changed to stop it in a specific case. It is usually eliminated by cut and try.

Single-stage poppet valves are the most liable to be unstable. This can be due either to their ability to react more quickly or because they have less damping than that of spool- or pilot-operated valves.

The fact that valve instability produces a pure tone shows that a resonance is involved. Attempts to explain these resonances in terms of the natural frequencies of their spring-mass systems have never been successful. These attempts even considered such things as the

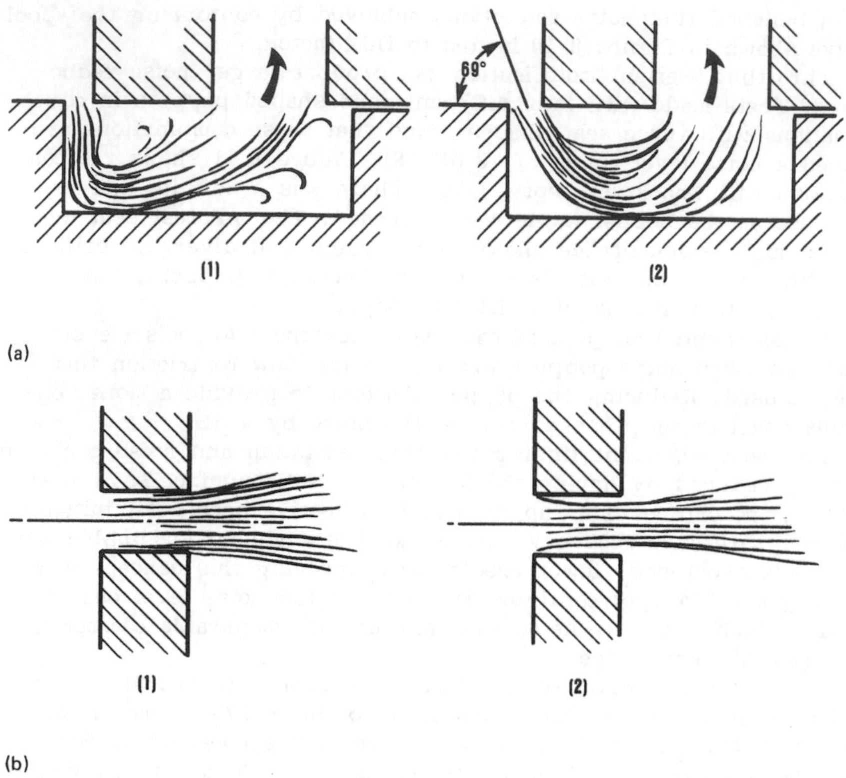

FIGURE 8.12 Bistable jets: (a) spool valve; (b) plate orifice.

mass of the fluid columns in attached lines and the compressibility of the fluid in the valve passages. All of these calculations assumed that the valve oscillates axially. Such resonances occur if the frequencies determined in this manner coincided with a pumping harmonic or are excited by a broadband noise source such as cavitation.

Research has shown that valves are capable of generating self-sustaining oscillations (11). One source of these oscillations is a flow instability called hydraulic jet flip. This takes several forms and occurs in many types of jets. Figure 8.12 shows how this occurs in a spool valve and in a plate orifice. Basically, it occurs because the jet may have two distinct forms, depending on its dimensions, size of its opening, and the pressure drop through it. At lower pressure drops, jets tend to cling to a wall, as in Figure 8.12(a1) and (b1). At higher pressures it takes on a more independent form, as in Figure 8.12(a2) and (b2). There is a transition pressure range where it exists in either form and may flip back and forth from one to the other. Each time it flips, the flow is perturbed and the valve begins to oscillate. Squeal from such sources occurs at some operating conditions and not others.

Flow and acceleration forces that tend to open the valve further than called for provide another motivation for valve oscillation. The onset of vibration is related to the strength of these forces in comparison to other parameters, such as stiffness, damping, and how much the ports open for a given valve movement. These forces increase with pressure drop through the valve, which explains why some valves provide good service at some pressures but begin to squeal at higher pressures.

The system in which the valve operates may have natural frequencies that are excited by transients and cause the valve to sustain the oscillations. This occurs with valves having just a little more than borderline stability. Valves with even greater inherent stability succumb if their natural frequency nearly matches that of the circuit. These valves operate satisfactorily in some circuits and squeal in others. Sometimes, changing the line connected to a valve makes this difference.

Line resonances, such as those discussed in Chapter 6, are a common source of such system-valve instabilities. Lines have a large number of resonant frequencies, which are determined by dividing their length by all odd integers to find the one-quarter wavelengths of these frequencies. Long lines increase the likelihood of an unfavorable match with a valve because these have a larger number of resonant frequencies in the range of valve natural frequencies. For that reason instabilities are sometimes avoided by using shorter lines. Valve inner chamber volume has an effect similar to adding line length, so reducing this volume is another option. Another way of discouraging this type of instability is to use lines composed of two different diameters so that reflections from the change in diameters interferes with the organ pipeline resonances (12).

Analyzing valves and their circuits to determine what factors must be changed to avoid instability requires considerable effort. Such effort is suitable for designing high-production-volume products, such as valves, but is generally impractical for developing quiet machines that will be built in relatively small numbers. As mentioned earlier, squeal problems in such machines are generally dealt with experimentally. Literature on this process universally prescribes replacing a squealing valve with one of a different design. Where the offender is a single-stage valve, it is best replaced with a pilot-operated valve.

Technicians sometimes stabilize squealing poppet valves by putting a permanent bend in their spring so that it rubs on the poppet stem and provides damping friction. Some chronic cases have also been helped by the addition of orifices in the valve passages. Unfortunately, these measures also degrade the valve's performance somewhat.

Pilot-operated valves have instabilities, although these are relatively rare. When it occurs after operating without trouble for a long time, it is usually cured by replacing worn seats or poppets.

One unstable poppet valve problem did not relate to most of the previous discussion. It was "cured" by installing guides that prevented sideways poppet motion. Such motion will produce flow perturbations as well as axial motions. This suggests that some valve instabilities are due to lateral vibration, which does not appear to have been considered in past research. The purposes of mentioning this case is to indicate that unorthodox experimentation is sometimes needed to solve a noise problem.

8.4 IMPACT NOISE

Impact noise is common but is seldom a problem. Most of it comes from solenoid valve actuators. Although it is possible to reduce it by using smaller clearances in the actuator linkage, this requires closer tolerances and makes the valves more expensive. The best way to reduce this noise is to isolate the valve as discussed in Chapter 10.

Another impact noise is worthy of mention here. It was a loud buzzing noise, at pumping frequency, that came from a large check valve. This is a poppet valve that offers very little restriction to flow in one direction but closes when the flow reverses. This particular valve had a small flow through it for a large part of the machine cycle. The small flow barely cracked the valve, and each pump flow pulsation caused the poppet to strike its seat. The cure in this case was to use a smaller valve that opened wider to pass the required small flow.

8.5 HYDRAULIC SHOCK

Hydraulic shocks are large pressure wavefronts that travel through a hydraulic system. Although they have the potential for causing serious mechanical failures, they rarely cause ruptures. However, they are considered to be a major source of system leakage because they cause hydraulic fittings to loosen. Of concern to us here is that they cause sudden, annoying loud noises.

Hydraulic shock occurs when valves are opened or closed rapidly. It is an old problem that occurs frequently in water distribution systems. Called *water hammer*, treatises on it date back to 1775 (13).

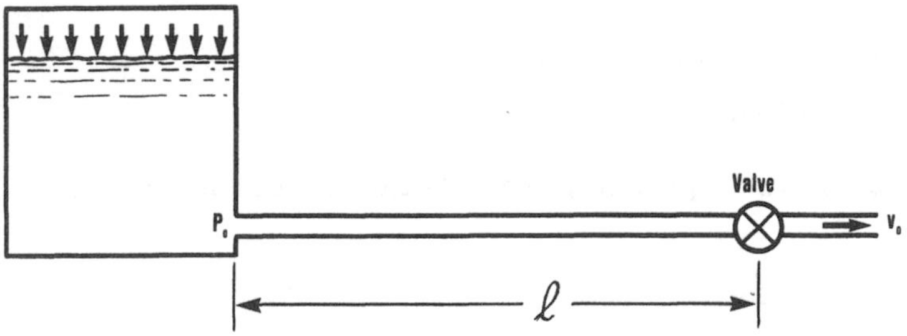

FIGURE 8.13 Water hammer demonstration circuit.

It is easiest to describe hydraulic shock by examining the mechanics of the simple circuit shown in Figure 8.13. Circuits like this rarely occur in machines; however, using a more practical circuit will needlessly confuse the discussion. Once the basic mechanism is understood, it should be easy to visualize it acting in practical systems.

Flow in the line shown in Figure 8.13 comes from a pressurized tank that provides a substantial line pressure. When the valve is closing, flow at the valve entrance slows and the pressure rises.

This is a classical example of kinetic energy being converted into potential energy. Potential energy is stored by compressing the fluid and dilating the line. For simplicity in this discussion, the latter factor is accounted for by suitably reducing the bulk modulus in calculating fluid compression. As with any spring, the energy stored in it is equal to one-half the product of the deflection times the resulting force. In this case the potential energy is

$$PE = \frac{\Delta p \ \Delta V}{2} \quad \text{in.-lb}$$

where

Δp = pressure increase, psi

ΔV = change in specific volume of fluid, in.

The bulk modulus relates the pressure and volume changes

$$\Delta p = B \frac{\Delta V}{V} \quad \text{psi}$$

where

B = effective bulk modulus, psi

V = specific volume of fluid, $in.^3$

If we combine the two equations, the potential energy is seen to be

$$PE = \frac{V}{B}\frac{\Delta p^2}{2} \quad \text{in.-lb}$$

The kinetic energy per unit weight of fluid is

$$KE = \frac{\Delta v^2}{2g} \quad \text{in.-lb}$$

where

v = flow velocity change, in./sec

g = acceleration due to gravity, in./sec^2

Equating the two energies yields

$$\Delta p = \Delta v \sqrt{\frac{B}{Vg}} \quad \text{psi}$$

Before we put this equation into a more useful form, it is well to note that the pressure rise is proportional to the square root of the bulk modulus. This term stands for the spring rate of the fluid path, not just the fluid. Therefore, system softness reduces the pressure rise.

Bulk modulus is equal to

$$B = c^2 \rho \quad \text{psi}$$

and

$$\rho = \frac{1}{Vg} \quad \text{lb-sec}^2/\text{in.}^4$$

where

B = effective bulk modulus, psi

c = speed of sound in fluid, in./sec

ρ = fluid mass density, lb-sec^2/in.^{4}

v = fluid specific volume, in.^{3}/lb

g = gravity constant, in./sec^2

applying these to the preceding equation leads it to reduce to

$$\Delta p = \rho c \Delta v \quad \text{psi}$$

If nothing interferes with the fluid deceleration process, all of the fluid velocity is converted into pressure

$$\Delta p = \rho c v \quad \text{psi}$$

This pressure adds to the line pressure supplied from the pressurized reservoir.

However, as the pressure rises at the valve face, the pressure is transmitted up the line toward the reservoir in the form of a pressure wave. This decelerates the flow in the rest of the line. At time T = l/c, the wavefront reaches the reservoir. This time is

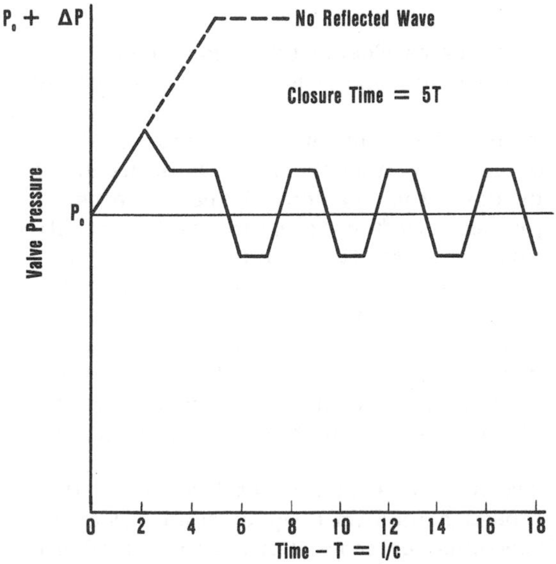

FIGURE 8.14 Valve pressure transients: slow closure.

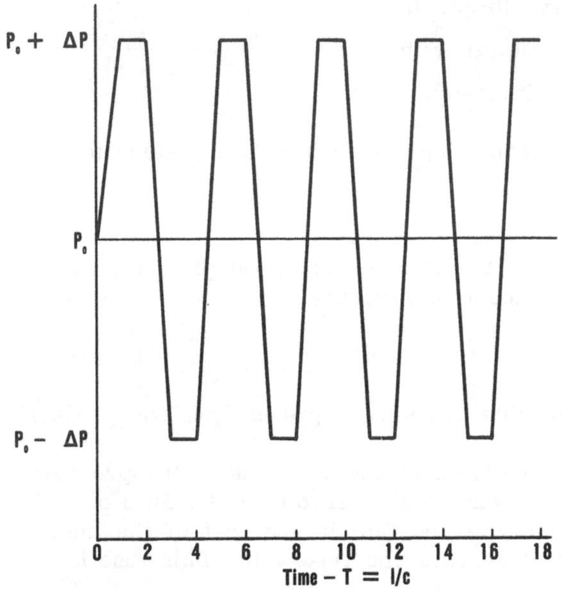

FIGURE 8.15 Valve pressure transients: fast closure.

measured from the start of the valve closing operation, l is the distance from the valve to the reservoir, and c is the speed of sound in the line.

As the wavefront passes into the reservoir, it "decelerates" the static fluid, so the fluid moves away from the line. This creates a decompression that proceeds down the line toward the valve as a negative wavefront. This process is often described as the negative reflection of the original wavefront. At 2T seconds the negative wavefront returns to the valve and begins to assist in decelerating the fluid so that the pressure stops rising.

Reference 14 is a classical testbook on water hammer that provides more detail on this mechanism and a graphical procedure for determining the pressure transients. The graphical procedure, together with equivalent algebraic and computer methods, is given in Ref. 15.

Figure 8.14 shows the pressure-time history at the valve. It was assumed that the closure caused a uniform drop in fluid velocity and that this closure takes 5T seconds for completion. The dashed line indicates the pressure profile that would occur if there were no wave action. It can be seen that the reflections are very helpful in attenuating the deceleration pressure rise.

If the valve closes very rapidly, reflections cannot attenuate the pressure rise. Figure 8.15 shows what happens when closure occurs in less than T seconds. The large pressure rise that occurs with such rapid valve closures is the hydraulic shock that we are concerned with here.

If the system has no losses, the wave reflections in this figure continue until another valve perturbation occurs. However, each time the wave passes through the line, it suffers friction and reflection losses. These losses cause the shock wave to damp out rapidly. It damps out even faster if its negative phase reduces line pressure below the fluid saturation pressure and cavitation occurs.

This shock also occurs when valves open rapidly. In this case fluid acceleration initially depresses the pressure, but when the resulting low-pressure wave is negatively reflected, it raises line pressure to about the same level as rapid valve closure, the only difference being that the wave is diminished by a line and a reflection loss before the high-pressure peak occurs.

Shock is not likely to occur in systems having short lines in which the wave return time is short. Its likelihood is also reduced by branches or changes in cross section that return reflections at different times. When shock occurs in such systems, the obvious way to eliminate it is to replace the valve with a slower-acting one. Another approach is to use flexible hose instead of steel lines. These decrease the effective bulk modulus of the fluid and reduce the shock pressure. Hoses should only be used when it is known that shock will occur, however. Since they also reduce the speed of sound, they also increase the reflection time and exacerbate borderline cases.

Shock in machines with long lines is recognized as a common problem, and manufacturers offer valves that are slow acting for use in such cases. Some of these valves have dashpots to slow valve movement. In spool valves, the lands are sometimes tapered so that greater travel is required to go from fully open to fully closed. In extreme cases, servo valves whose motions can be carefully controlled are used.

Accumulators behave the same as the pressure-loaded reservoir in the earlier discussion, in that they cause negative wave reflections. They can therefore be connected to long lines, near the valve, to shorten the line acoustically. They can also be viewed as making the system much softer so that it absorbs the kinetic energy with less pressure rise. Since their storage capacity is so great in comparison to a line, even small accumulators are effective in reducing valve-induced shock.

REFERENCES

1. R. A. Heron and I. Hansford, "The Development of a Method for Measuring the Sound Power Generated by Oil Hydraulic Relief Valves," paper C385/80, *Proceedings of the Quieter Oil Hydraulics Seminar,* Institution of Mechanical Engineers, Oct. 1980, pp. 77–85.

2. R. Heron, "Valve Noise," *Quieter Fluid Power Handbook,* BHRA, The Fluid Engineering Centre, Bedford, England, Chap. 5, pp. 23–27.

3. A. E. Moore, "Oil Hydraulic Cavitation Noise from Simple Flow Restrictions," interim report BH6, contract no. K/A72b/136/CB/A72b, BHRA, The Fluid Engineering Centre, Bedford, England, May 1978.

4. A. Crook and I. Hansford, "Oil Hydraulic Relief Valve Noise," paper C261/77, *Proceedings of the Quiet Oil Hydraulic Systems Seminar,* Institution of Mechanical Engineers, Nov. 1977, pp. 59–67.

5. R. E. Hatton, *Introduction to Hydraulic Fluids,* Reinhold, New York, 1962, p. 35.

6. J. Kane, T. D. Richmond, and D. N. Robb, "Noise in Hydrostatic Systems and Its Suppression," *Proceedings of the Institute of Mechanical Engineers,* Part 3L, 180: 1965–1966.

7. C. Donaldson, "A Quiet Valve," *Quieter Fluid Power Handbook,* BHRA, The Fluid Engineering Centre, Bedford, England, 1980, Chap. 12, pp. 86–94.

8. R. A. Heron and I. Hansford, "Reduction of Relief Valve Noise by Detail Design of the Restriction," added paper, *Proceedings of the Quieter Oil Hydraulics Seminar,* Institution of Mechanical Engineers, Oct. 1980.

9. J. E. Funk, "Poppet Valve Stability," paper no. 62-WA-160, American Society of Mechanical Engineers, 1962.

10. H. E. Merritt, *Hydraulic Control Systems,* Wiley, New York, 1967.

11. J. F. Blackburn, J. L. Coakley, and F. D. Ezekiel, "Transient Forces and Valve Stability, in *Fluid Power Control,* J. F. Blackburn, G. Reethof, and J. L. Shearer, eds., Wiley, New York, 1960, Chap. 12, pp. 359–400.

12. Ref. 11, p. 440.

13. L. Dodge, "Reduce Fluid Hammer," *Product Engineering,* pp. 88–95, Dec. 10, 1962.

14. L. Bergeron, *Water Hammer in Hydraulics and Wave Surges in Electricity*, Wiley, New York, 1961.

15. E. B. Wylie and V. L. Streeter, *Fluid Transients*, McGraw-Hill, New York, 1978.

9
Machine Airborne Noise

Hydraulic machine noise is an energy chain. It starts with pumps, motors, and valves generating acoustic energy, only a small part of which is sound. The rest of this energy is in the form of fluid pulsations and vibrations which are transmitted through various paths to other components throughout the machine. Some of these components also radiate sound. Our concern is only with sound that is heard, so the last link in this energy chain is the sound radiated to a listener. Noise reduction is achieved by breaking or weakening the links in this chain.

In this chapter we deal with interdicting the last link in the chain. In an industrial situation the listener is generally the machine operator, but the possibility that it is a person in another nearby workstation or passing by the machine must also be considered. The object of this chapter is to reduce the noise reaching the usually well-defined locations of these listeners.

9.1 DISTANCE

This noise control area sometimes affords quieting at very low cost. Sound pressure decreases by the square of the distance it travels. This means that the sound pressure level drops 6 dB each time the distance from the noise source doubles.

Perhaps a homely example is in order. A machine had its hydraulic power supply located in back of the machine head, about 5 ft from the floor. It was therefore only about 2 ft from the operator's head. Moving the power supply to the floor increased the distance from the operator to 6 ft. Since sound pressure level is inversely proportional to distance, this move reduced the operator's noise level by about 9 dB, a very worthwhile improvement.

It can be argued that such obvious opportunities are unlikely. Although this is true, it did happen and it should encourage us to look for similar, perhaps not quite as simple options. For example, quite commonly radiators can be located so that large components, such as reservoirs, shield listeners from their noise. This makes the sound travel around the shield to reach the listener. Moving the operator's controls to a quieter location is another possibility.

Relocating radiators or people does not reduce noise energy radiated by a machine. Such changes may make a machine meet workplace OSHA limits, however. This is pointed out to emphasize the need to set quieting goals in terms of noise reaching people. Although most noise reductions involve reducing noise radiation, valuable opportunities are overlooked when only this option is considered.

9.2 RADIATION SOURCES

Reducing machine noise by modifying radiation sources is usually not given the attention it deserves. The general characteristics of some of the common radiators are discussed in Section 4.1. Here we look at how these contribute to machine noise and how their effectiveness is reduced.

9.2.1 Structural Members

Machine structural design is an area in which dramatic reductions in noise are frequently possible. Noise control lecturers often demonstrate the importance of good radiators by showing how the noise of an electric buzzer is increased by placing it on a tabletop. Machine designers do not mount hydraulic noise sources on drumheads, but many are not aware of the radiation efficiency of some of the structural members that they do use. The energy source does not have to be mounted directly to such members to cause radiation; vibratory energy once introduced into a welded structure is efficiently transmitted to all the structural members.

Lightweight and wide-flange I beams are popular with machine designers because they have good overall stiffness for their weight and broad surfaces for mounting components. They are also excellent noise radiators, especially in the frequencies above 500 Hz. Angle iron, which is also frequently used, is another good noise radiator.

Lateral deflections of these members are important only for lower frequencies. Higher vibration modes, like that shown in Figure 9.1 for an I beam, are responsible for radiations at the more critical middle and high frequencies. Resistance to such vibration is greatly enhanced by occasional welded steel straps that tie the flanges together, as shown in the same figure. Stiffeners can also be applied to angles to reduce their radiations.

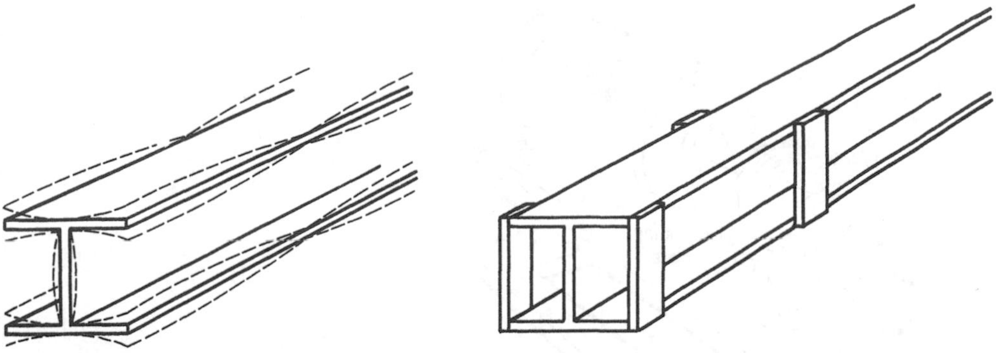

FIGURE 9.1 I beams are good noise radiators.

Stiffeners added to structural members are poor in appearance and are an added expense. They are not needed when square mechanical tubing, like that shown in Figure 9.2, is used. This shape also has a very neat appearance. Box shapes formed by welding channels together, as shown in Figure 9.2, also make neat low-radiation structural members. Only intermittent welds are required because they are lightly stressed.

Plates used in machine structures are some of the best sound radiators. They have so many vibration modes that they usually have resonant frequencies coinciding with or are close to strong pump harmonics. The best strategy for minimizing radiation from plates is to isolate them mechanically from the noise energy source. When this is not possible, the plate should be made as small as practical.

Another way to reduce plate radiation is to cut out the center of the plate. Figure 9.3 shows a pump and motor assembly, typically

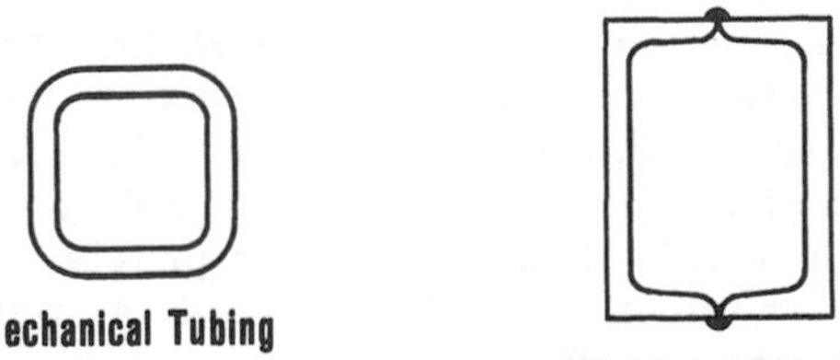

FIGURE 9.2 Preferred low-radiation structural members.

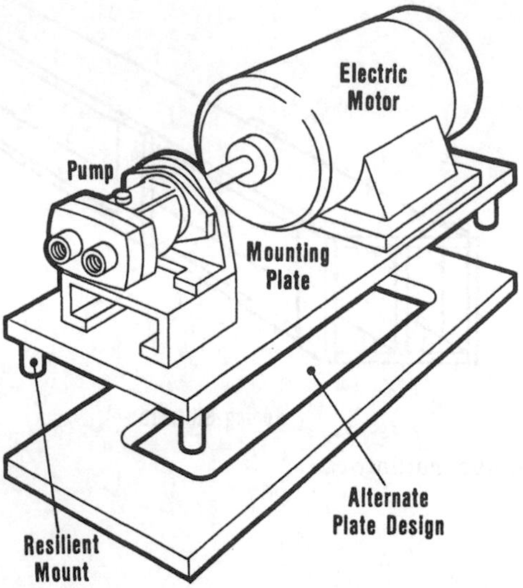

FIGURE 9.3 Cutting holes in thick plates reduces noise radiation.

mounted on a plate. The plate is generally thick to avoid excessive deflections and mixalignment due to pump and motor weights and shaft torque. A surprising 12-dB reduction in the noise radiating from an assembly of this type occurred when the center of the plate was cut out as shown.

Some of the noise reduction from cutting out the center of the plate obviously occurred because the air could then "slosh" from the top face to the bottom as the plate remainder vibrated. The effect was like removing the baffle from around a radiating piston as shown in Figure 4.8. From this figure, however, it can be seen that this factor does not fully explain the reduction that occurred.

The result was especially surprising because, as discussed in Section 4.1.6, sound radiating from a plate is generally from the edges and corners, as shown in Figures 4.11 and 4.12. Radiation from the central portion is in the form of quadrupoles which are very inefficient, so the removal of this portion would not be expected to have a large impact. However, the plate was 1 in. thick and, as seen in Figure 4.13, had a critical frequency slightly less than 500 Hz. Because of this the maximum plate deflection points were too widely spaced to act as quadrupoles above this frequency. The plate's radiation efficiency, like that of the 1-in. plate in Figure 4.14, was high for the strongest pump harmonics. The plate center, therefore, had been contributing a large portion of the noise.

Increasing plate thickness appears to be a good way to reduce plate radiation. Plate stiffness is proportional to the cube of thickness, so doubling thickness reduces deflections of stiffness-controlled frequencies by a factor of 8. This would be expected to reduce noise radiation at these frequencies by 18 dB. Vibration of mass-controlled frequencies would be reduced only 6 dB. However, doubling thickness also cuts the plate critical frequency in half. As seen in Figure 4.14, for plates 1/4 in. or less, this has a big effect on the radiation efficiency in the important range 500 to 2000 Hz. This may or may not outweight the effect of reducing deflections. To generalize, increasing plate thickness does not always reduce noise. Its effectiveness in a particular set of parameters is difficult to calculate because of the many frequencies and modes that must be considered. For this reason it is thought best that it be evaluated by trial.

9.2.2 Lines and Hoses

Hydraulic lines are a leading airborne noise source in machines. In pump rating tests, for example, it was found that unenclosed discharge lines usually produced more noise than the pump itself. The author also recalls a noisy machine that had its power supply in a very remote location. The supply line radiated enough noise along its long path to make it clearly heard over all other noise sources throughout the plant.

As discussed in Section 4.1.5, lines respond to two forms of noise energy. Where all solid lines are used, structureborne noise-induced radiation predominates, even though it is less efficiently radiated than fluidborne-induced radiation. This is due to the very high level of vibratory energy received directly from the pump.

Flexible hose has practically no lateral stiffness, so it does not respond to pump vibration. This is discussed in detail in Section 10.6.1. Where hose is used instead of solid tubing or pipe, fluidborne noise-induced radiation predominates. The amount of noise radiated depends on the type of hose used.

A good comparison of these different types of sound radiation is possible from data reported by BHRA Engineering (1). A gear pump was operated with a number of different discharge lines. Each was 10 m long and was terminated by a valve. Airborne noise radiated from the central 3.5 m was measured. A smoothing technique was applied to the data to eliminate the effects of the many resonances that occurred in the testing. Only the 1800-rev/min data are used here.

The sound levels from two identical 1-in. steel tubes, one directly attached to the pump and valve and the the other isolated from these components with hose, are shown in Figure 9.4(a). The investigators

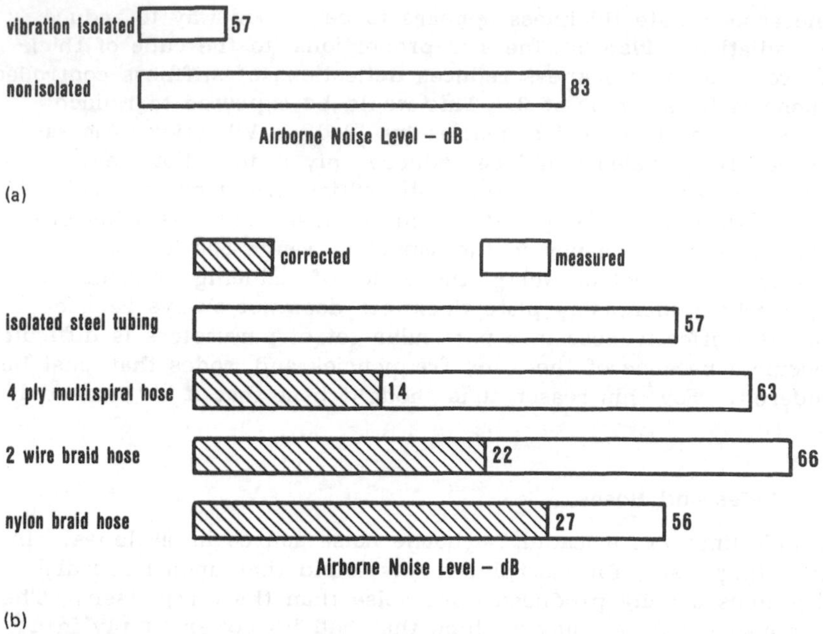

FIGURE 9.4 Hydraulic-line airborne noise radiation: (a) effect of vibration isolation of solid lines; (b) comparison of fluidborne noise-induced radiation by various types of hydraulic lines.

noted that 1 m of the nonisolated line radiated as much noise as the pump itself, establishing that the line was a major noise source. Since this line produced 26 dB more sound than the isolated line, the comparison clearly demonstrates that lateral bending is responsible.

In hose tests, the fluidborne noise in the line varied with the material used. It is assumed that the airborne noise is due to these pulsations, so the data were normalized to eliminate the effect of this difference. This was done by subtracting the fluidborne from the airborne levels. Since both levels are in logarithmic units, decibels, this has the effect of dividing by the fluidborne level. The results were negative numbers, so the corrected radiation level from the isolated solid tube, which was the lowest, was used for a baseline and the amount needed to make this statistic zero was added to all corrected levels.

The measured and corrected levels are shown in Figure 9.4(b). All tests were run at 3000 psi, with the exception of the nylon braided hose test, which had to be limited to 1350 psi because of its lower strength.

The hoses radiated far more noise than did the isolated steel line. Nylon hose, which has the least resistance to pressure dilation, produced the most. It was surprising to note that braided hose produced 8 dB more than spiral-wrapped hose. Although some of this may be attributable to differences in diametral stiffness, it probably indicates that the braided hose was also vibrating laterally. As discussed in Section 10.6.1, this hose is only 40% as effective as wire-wrapped hose in attenuating structureborne noise.

Since hose is a good radiator, long lengths should be avoided. The quietest long hydraulic lines consist of solid tubing or pipe with relatively short lengths of hose at each end. Such lines utilize the vibration attenuation of hose and the low pulse radiation of solid line.

Pressure pulsations generate line vibrations whenever the line contains a Z or U configuration (2) as shown in Figure 9.5. Pressures at the ends of the offsets in these configurations are out of of phase with each other by virtue of the distance between them. This phase difference, for a given fluidborne frequency, increases as the offset length approaches an odd multiple of the half-wavelength and decreases as the length approaches a multiple of the wavelength. Determination of this wavelength is discussed in Section 6.1.1.

The phase differences at the two ends of the offset provide a forcing function that periodically accelerates the liquid in this portion of the line. This action produces a periodic inertia force, at right angles to the rest of the line, which excites the vibration.

The amplitude of this vibration depends not only on the phase difference between the offset ends but also on the relation between the excitation frequency and the lines transverse resonant frequencies. Because of the bends, conventional natural frequency calculations are difficult. However, studies show that a reasonable approximation is easily made by assuming that the line is straight (2).

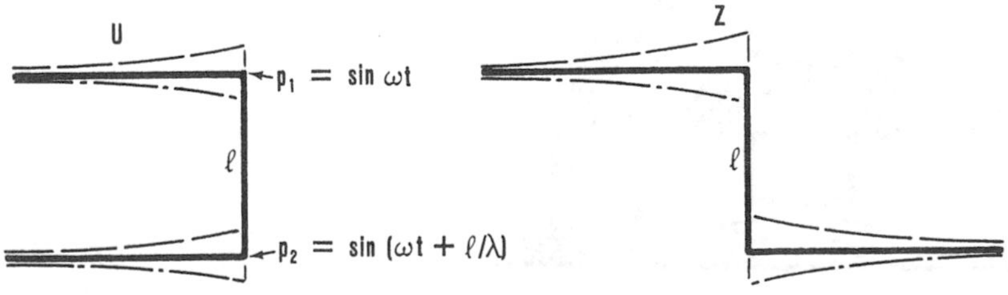

FIGURE 9.5 Line vibrations result from U and Z offsets. (From Ref. 2.)

9.3 ENCLOSURES

Enclosing noise radiators so that their sound reaching listeners is greatly attenuated is a frequently used control. Where existing structure can be used, this can be the cheapest way to reduce a machine's noise.

Enclosures, in general, have a bad reputation. Many are unsightly and look like the afterthought that they are. Some are cumbersome and make machine operations or servicing more difficult. Such enclosures are usually removed after a short time in service. This reputation is the result of poor engineering. There are also many examples that show that enclosures can be attractive as well as cost-effective when well engineered. These include familiar examples such as the inside of your car, outboard engines, and kitchen sink disposals.

Designing good enclosures is a real challenge. They not only must have low cost, good appearance, and accessibility, but they must be installed while accommodating lines and shafts that must pass through their walls. Adding to this, most suitable materials are available only in sheets and panels. As we will soon discuss, most of the enclosure fittings are not on the market, and providing them adds to the design effort.

9.3.1 Enclosure Principles

Enclosures are simple devices but have rigid requirements. When one of these is ignored, the effectiveness of the enclosure is curtailed severely. The four essential elements of enclosures are shown in Figure 9.6.

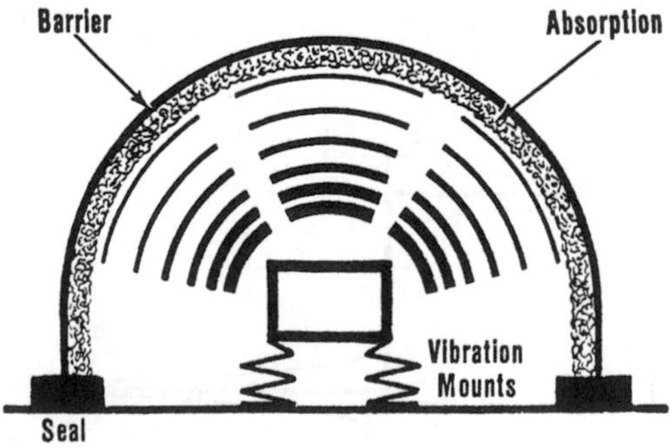

FIGURE 9.6 Enclosures must meet all four requirements to be effective. (From Ref. 14.)

One of the requirements is that the enclosure be mechanically isolated from the noise source. Because the enclosure is larger than the source, it tends to be a better radiator. Without isolation, an enclosure may increase noise rather than attenuate it.

Isolating hydraulic noise sources utilizes well-established technology that is discussed in Chapter 10. Hydraulic lines attached to these sources are also capable of transmitting noise energy to an enclosure, and their isolation is discussed in detail in the following section.

Sound impinging on an enclosure wall is reflected mostly as shown in Figure 9.7. This illustrates the path of a "ray" of sound as it reflects within the enclosure. A small amount of the sound, depending on the wall material and the angle of incidence, passes through the wall at each reflection. The sound intensity level in the enclosure is the total effect of all of these sound rays. The sound passing through the walls is a fixed percentage of this level. The difference between the level inside and that just outside, in decibels, is the *transmission loss.*

When the sound energy passing through the walls is not equal to the amount being generated within the enclosure, the sound level at the inside surfaces rises until it does. With a wall material providing a transmission loss of 30 dB, the average sound pressure on the enclosure's inside surfaces will rise by this amount if there are no other losses. Then when the elevated sound level is attenuated by 30 dB, the sound outside the enclosure is the same as if the enclosure was not there.

This assumes that none of the sound energy is transformed into heat. Fortunately, some sound energy is converted to heat by friction in the air and by the reflection process. As the walls vibrate in response to the sound, hysteresis within the wall material also converts some acoustic energy to heat. The energy reaching the outside of the enclosure as sound is reduced by these conversions. This reduction, although appreciable, is usually small in comparison to the transmission-loss potential of the walls, however.

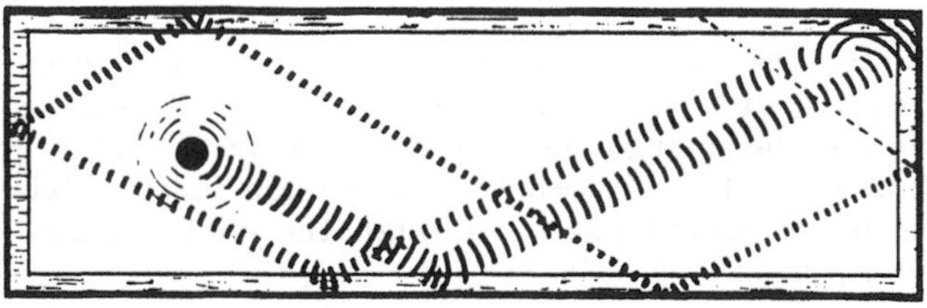

FIGURE 9.7 Sound ray reflection in an enclosure. (From Ref. 13.)

To be fully effective, an enclosure must be lined with sound-absorptive material. In Figure 9.7 the sound ray is diminished because it makes two passages through an absorptive lining, each time it is reflected. The sound pressure level exerted by all such reflections, then, is much lower than if absorption did not occur. The intensity of the sound reaching the outside, being a percentage of the inside sound level, is also reduced. The noise reduction afforded by the enclosure, then, is the transmission loss provided by the enclosure itself, minus the sound pressure rise in the enclosure. This attenuation is called the *insertion loss* of the enclosure.

The sound pressure level produced by a sound source within an enclosure is

$$L_p = 10 \log \frac{W\rho c}{400} \left(\frac{1}{4r^2} + \frac{4}{R} \right) + 10.8 \quad \text{dB}$$

where

W = sound source, W

ρc = characteristic resistance of air, 406 N-sec/m^3

r = distance from source to measurement point, ft

R = room constant, ft^2

The first term in the parentheses is for the sound pressure coming directly from the source. It is the pressure that occurs if the enclosure is not there. The level of this pressure drops 6 dB each time the distance r is doubled. The second term is the pressure level contributed by reflected or *reverberant* sound. This pressure is constant throughout the enclosure. At some distance from the source, depending on the sound energy losses, the reverberant pressure equals the direct pressure. At lesser distances direct sound predominates, and at greater distances reverberant sound predominates.

This equation includes both SI and English units. The last term reconciles these units. If meters are used instead of feet, this term is omitted.

The room constant is a measure of sound energy losses in the enclosure. These losses are due primarily to absorption at the wall surfaces but also include the energy passing into the walls. If large rooms were being considered here, it would also include viscous friction losses occurring as the sound passes through the enclosed air. The room constant is equal to

$$R = \frac{S}{1 - \alpha} \quad \text{ft}^2$$

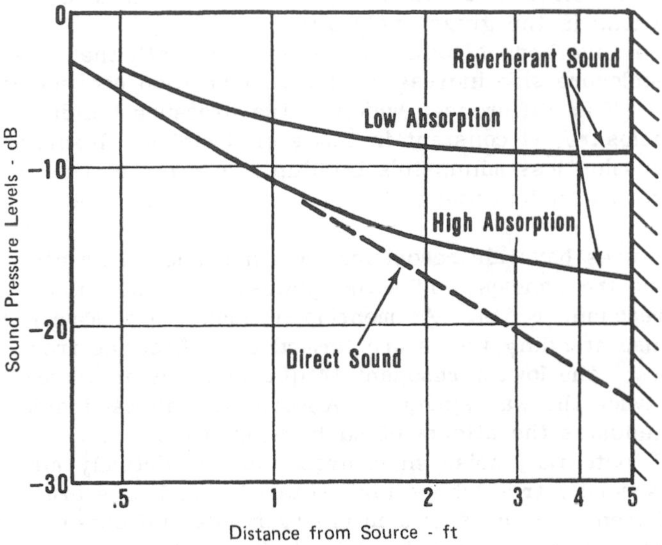

FIGURE 9.8 Sound pressures along the axes of a 10-ft^3 room with the sound source at the center.

where

S = total enclosure internal surface area, ft^2

α = arithmetic average absorption coefficient

The relationship of the pressure components within an enclosure is shown in Figure 9.8. This example assumes that the sound source is small, does not emit any discrete frequencies, and radiates uniformly in all directions. It also assumes that the reverberant field within the cubic enclosure, 10 ft on a side, is diffused. Hydraulic noise, of course, has discrete frequencies and it is impossible to have a truly diffused field in an enclosure with parallel sides. These two factors produce standing waves in practical enclosures. The calculated pressures are therefore only averages of pressures along the central axes of the enclosure.

The six sides of the 10-ft cubic enclosure have a total surface area of 600 ft^2. Two absorption coefficients were used for this example. The first, 0.05, provided a room constant of 31.6 ft. The second, 0.80, provided a room constant of 240 ft. Absorption coefficients are discussed in detail in Section 9.3.4, so the only comment made here is that most coefficients fall within this range. The calculated pressures therefore illustrate fairly well the extremes in pressure distributions in the given enclosure.

Both direct and reverberant sound levels decrease with the square of distance, as enclosure size increases, if the absorption coefficient remains the same. This difference, which is the pressure buildup caused by the enclosure, is constant in terms of decibles. Insertion loss is the transmission loss minus this buildup. Therefore, the enclosure effectiveness is independent of size as long as the same materials are used.

Practical enclosures have air resonances which reduce their effectiveness at some frequencies. Like flat plates, they have a galaxy of these resonant modes. As mentioned earlier in this section, these occur because standing waves are formed by reflections from their parallel sides: the lowest resonant frequencies having wavelengths equal to twice the wall spacing. Absorptive material inside the enclosure extenuates the effects of such resonances.

To achieve full potential, noise enclosures must be virtually airtight. This is especially true where high transmission losses are sought. The difference in the sound intensity inside and outside the enclosure is so great that even small cracks leak significant amount of sound. Figure 9.9 shows how the transmission loss is reduced by small openings. From this figure we see that if an enclosure has openings equal to 1% of its surface area, it is impossible to achieve a transmission loss above 20 dB.

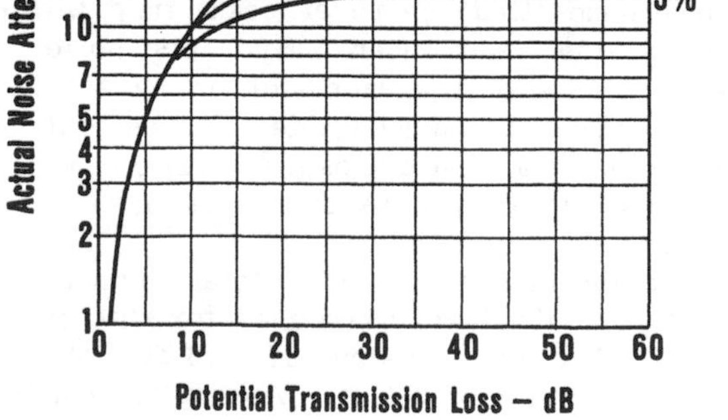

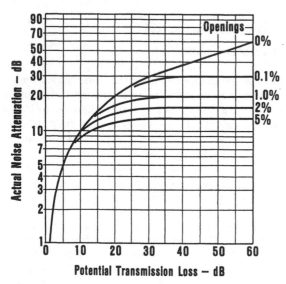

FIGURE 9.9 Openings reduce enclosure noise reduction. (From Ref. 14.)

Enclosures are made in parts to facilitate their installation. Joint lines tend to be long, and even small gaps along their length lets significant noise escape. All joints, must therefore, be gasketed and tightly clamped to prevent such losses. Where access is required for servicing, panels should be secured with toggle or other rapid-action clamps to provide a tight seal while making removal quick and easy.

9.3.2 Enclosure Access

It is particularly difficult to provide tight enclosures for hydraulic equipment because access must be provided for lines and shafts.

Hydraulic lines vibrate and, as discussed earlier, must not touch the enclosure walls. The difficulty with lines passing through enclosure walls is that they are not precisely located related to the enclosure. The gap must therefore, be made larger than required for vibration isolation. From a noise standpoint, clearance around these lines must be as small as enclosure and hydraulic assembly tolerances permit.

If only 10 to 12 dB of transmission loss is needed, the gaps between carefully sized holes and lines may not be a problem. For higher transmission losses, the gaps must be filled.

In laboratory setups, gaps can be filled with nonhardening mastic or putty, as shown in Figure 9.10, without transmitting vibration to the wall. Its density and thickness determine its effectiveness. Unfortunately, putty barriers are not suitable for production machinery.

Closed cell sponge or rubber grommets, like that also shown in Figure 9.10, are used to fill the gaps on production machines. To my knowledge these are not available commercially, so they must be made on an ad hoc basis.

Where high transmission losses are required, devices that adjust for variations in line location without leaving a large window for sound to escape must be used. Figure 9.10 shows a device that does this. It fits tight on the line and is fastened to the enclosure wall. A resilient sleeve prevents energy transfer to the wall. Unfortunately, devices of this type are not available commercially and have to be fabricated by the machine builder.

Pump and motor shafts cause the same problems as hydraulic lines. However, since they do not vibrate, clearance around them can be smaller. They do have the problem of location relative to the enclosure so their closure devices must be able to adjust for position variations. Sometimes they also need to be split longitudially, so that the two halves can be installed without removing the shaft coupling.

The line clearance closure devices, shown in Figure 9.10, are not suitable for shafts because they rub and cause friction heat. Because contact with the shaft is not permissible, no closure can eliminate the gap altogether. The strategy, then, is to use two widely

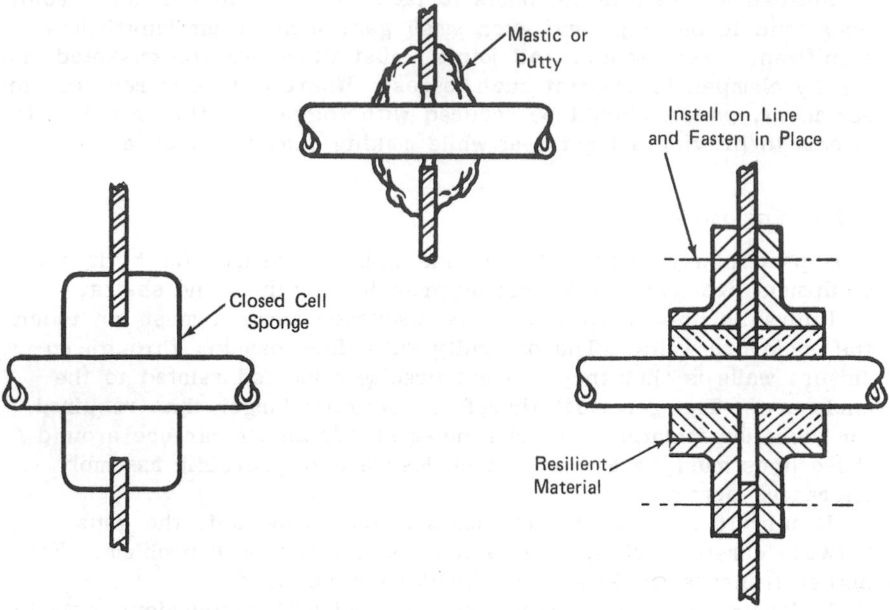

FIGURE 9.10 Line clearance closures.

separated gaps in series. Their net effect is that of a very small gap. Further reduction in the noise escaping through the gaps can be attained by adding sound absorption in the space between the two gaps. Two shaft noise locks using these concepts are shown in Figure 9.11. Again, these devices are not commercially available and offer a challenge to the machine designer.

It sometimes seems preferable to enclose both a pump and its drive motor. This eliminates the shaft problem and attentuates the noise radiated by the motor. The problem then becomes one of providing sufficient cooling. Pumps, unless they operate for long periods near cutoff and high pressure, are cooled by the fluid they circulate. The electric motor, however, is air cooled and needs a large air supply. Figure 9.12 shows the way that this air is generally provided.

Sound escaping through the necessarily large openings in the chamber enclosing the motor-pump assembly is absorbed as it passes through the length of the ducts. This means that the ducts must be fairly long and, of course, lined with sound-absorbing material. The design of such ducts is discussed in a Section 9.3.6. Unfortunately, the duct features that promote absorption also restrict airflow, so a fan must be added to the enclosure to overcome the added resistance.

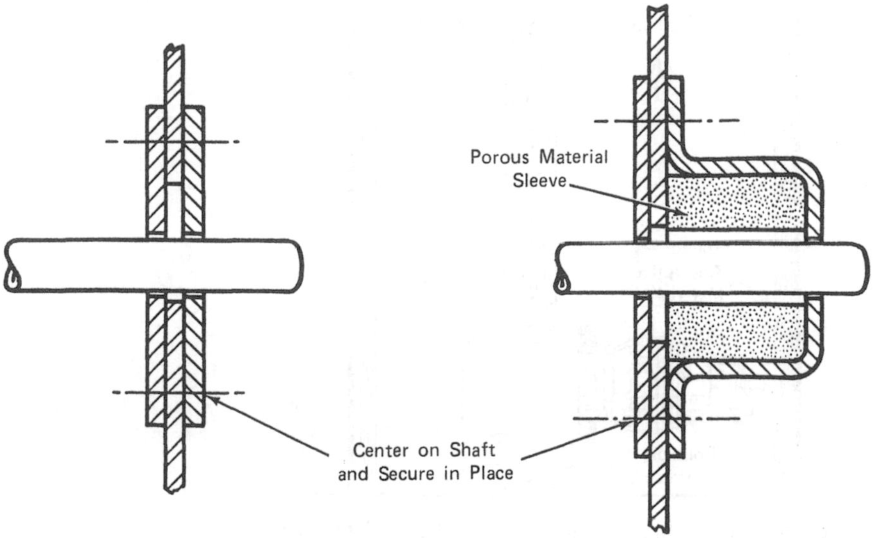

FIGURE 9.11 Shaft noise locks.

The fan in the electric motor can be used to provide sufficient cooling air if the ducts are made large enough. Figure 9.13 shows an arrangement of this type. This scheme was used for a 100-hp power supply in a noise test facility which had good height and noise absorption. Although the ducts provided low attentuation per foot, they were sufficiently long to meet their purpose. Because the room walls and ceiling were covered with 4-in. fiberglass batts, very little of the noise escaping the stacks was reflected toward the place where noise measurements were being made.

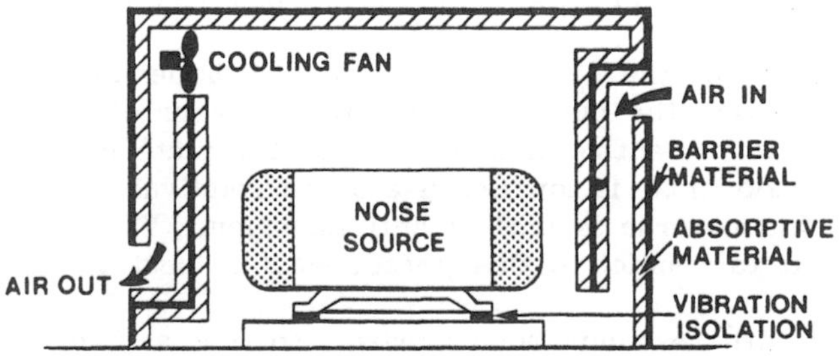

FIGURE 9.12 Air-cooled motor-pump enclosure. (From Ref. 13.)

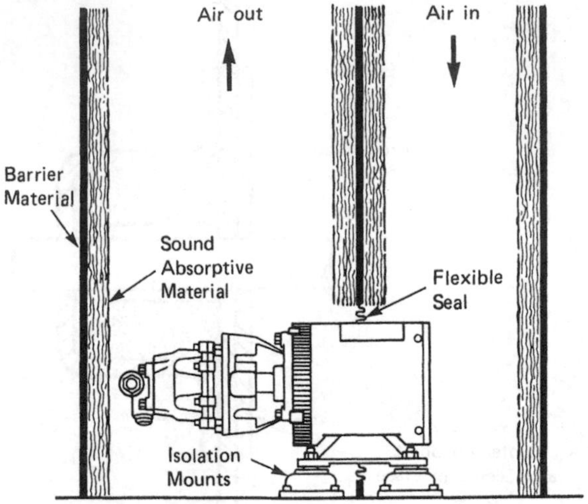

FIGURE 9.13 Stack enclosure. (From Ref. 14.)

9.3.3 Barrier Materials

Enclosures are made of barrier materials that are impervious to air. It is generally said that the best materials are heavy and limp. This is an oversimplification, since the transmission loss provided by barrier materials is due to a number of distinct mechanisms, each predominating over a frequency range.

The transmission loss of a barrier material panel varies with frequency as shown in Figure 9.14. At low frequencies, the panel responds like a diaphragm to sound pressure impinging on its receiving surface. In terms of the discussion in Section 3.1.4, it vibrates in the (1,1) plate mode. At these frequencies it is stiffness controlled; that is, the transmission loss is proportional to the plate stiffness and it decreases proportionally with frequency (−6 dB/octave).

The next phase starts as the (1,1) mode resonant frequency is approached. In this region the plate responds to a number of resonances and, in between, antiresonances. When the plate is resonating its transmission loss is low because of its relatively high vibration velocity. The reverse is true at antiresonance. Transmission loss is difficult to predict in this phase, so it is best to avoid it.

Damping increases with resonant mode number, so that at two to three times the (1,1) resonant frequency, the peaks and valleys in the transmission loss curve are small enough to ignore and the

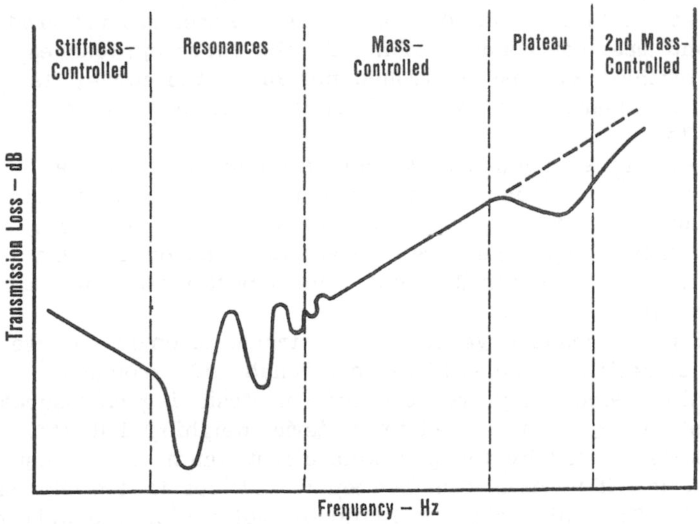

FIGURE 9.14 Enclosure panel transmission loss.

next phase starts. In this range, transmission loss is proportional to both plate mass per unit area and frequency (+6 dB/octave). It is generally referred to as the mass-controlled region.

The coincidence effect controls the next phase, which starts as the plate critical frequency is approached. Coincidence is where the sound and plate bending wavelengths are equal and a sort of resonance occurs. As discussed in Section 4.1.6, plates become very good radiators at the critical frequency.

Coincidence also makes the plate very responsive to noise at this frequency. As a result, transmission loss dips in the region of the critical frequency. This dip is relative to the mass-controlled portion of the curve, so in absolute terms it appears as a shallow horizontal wave. The exact shape of the curve in this region depends on damping, both in the barrier material and at joints. Generally, the curve is somewhat sinusoidal, with a total range of around 8 dB. Where high damping occurs, the curve becomes almost flat. Damping also determines the width of the region. This region, which is called the plateau, normally extends from about 0.3 to several times the critical frequency.

The transmission loss in the plateau region is a function of both the critical frequency and the relative location of the mass-controlled segment of the curve. For a given material, both of these are proportional to barrier thickness. The plateau transmission losses, then, remain the same for all thicknesses of a given material.

After the plateau, transmission loss is again related to unit mass but increases at a rate of 10 dB octave. This is sometimes referred to as the second mass-controlled region. As the curve approaches the extension of the first mass-controlled portion of the curve, its slope decreases. When it is within 5 dB of the extension, it becomes parallel to it.

In considering transmission loss we only need to focus on the frequency range where hydraulic noise is generally highest, in the range 500 to 2000 Hz. If adequate loss is provided in this range, losses at other noise frequencies are also apt to be adequate. On this basis, only the mass-controlled and plateau portions of the curve are relevant.

Figure 9.15 is an idealized version of the transmission-loss curve that is useful in design. The stiffness-controlled and resonance regions are omitted since they are not useful in designing enclosures. The abscissa of this curve is scaled for material weighing 1 lb/ft^2. When this scale is divided by the per square foot weight of a panel of a given material, it becomes the frequency scale for that thickness of that material. The unit weight is found by multiplying the weight of a square foot of a 1-in.-thick plate of the material, given in Table 9.1, by the thickness.

The mass-controlled line applies to all materials, up to their plateaus. The plateaus are shown as horizontal lines, with the appropriate height and width for a number of common enclosure materials (3). These are adequate approximations for many applications; however, when they are used it should be remembered that actual transmission losses might be from 5 to 7 dB lower at the critical frequency.

The plateaus in Figure 9.15 have two dots shown on them, one near the center and one near the right end. These correspond to the one-quarter and three-quarter frequencies of the plateau range and were added to make it easy to make a better estimate of the plateau transmission losses. This is done by drawing in a curve that rises 3 dB at the first dot and falls to 5 dB below the line at the second dot (3), as shown by the dashed curve on the plywood plateau.

Thickness determines the frequency where a panel of a given material reaches its transmission-loss plateau. The thicker the material, the lower the frequency at which the plateau is reached. Where the required loss in the critical range 500 to 2000 Hz is nearly equal to the plateau height, sufficient thickness must be used to ensure that the plateau is reached at a frequency below 500 Hz. The benchmark thicknesses needed to meet this condition, for the materials covered by the design chart, are given in Table 9.1. For greater thicknesses, the transmission losses are equal to the plateau values, also given in this table. Losses for thinner panels are determined from the mass-controlled line of Figure 9.15, up to their plateau limit.

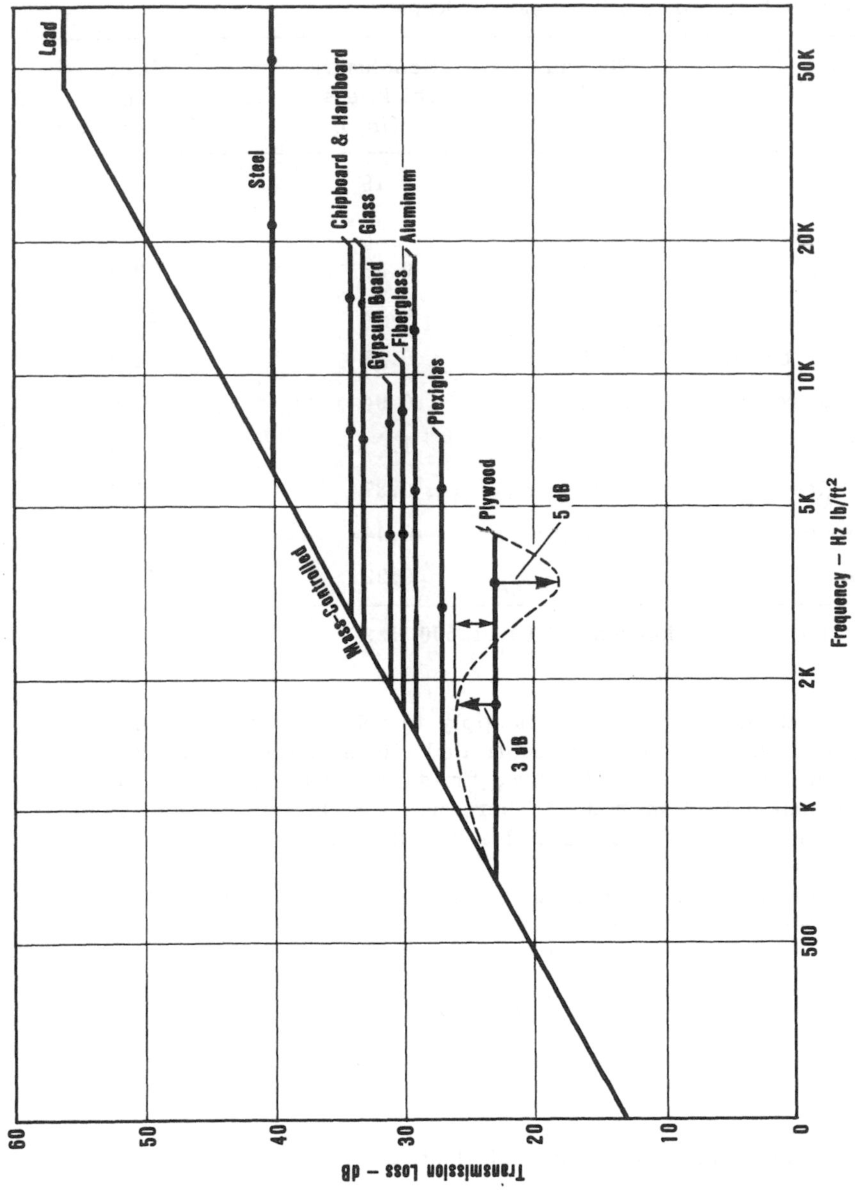

FIGURE 9.15 Barrier transmission-loss design curve.

TABLE 9.1 Enclosure Barrier Thickness[a]

Material	Plateau TL (dB)	Benchmark thickness (in.)	Unit weight ($lb/ft^2/in.$)
Aluminum	29	0.19	14.2
Chipboard	34	1.42	3.3
Fiberglass	30	0.33	9.0
Glass	33	0.32	13.0
Gypsum board	31	0.83	4.0
Hardboard	34	0.94	5.0
Lead	56	1.	59.0
Plexiglas	27	0.37	5.6
Plywood	23	0.44	3.0
Steel	40	0.22	41.8

[a]Plateau transmission loss (TL) at 500 Hz.

This design curve does not apply to relatively small panels. As discussed earlier, the mass-controlled phase of the curve starts at a frequency that is two to three times the resonant frequency of the (1,1) plate vibration mode. Therefore, for the curve to be valid in the critical frequency range, this resonant frequency must occur below 500/3 = 167 Hz.

From the equation given in Section 3.1.4, it can be seen that this resonant frequency is

$$f_{1,1} = 368t \sqrt{\frac{E}{W}} \left(\frac{1}{a^2} + \frac{1}{b^2} \right) \quad \text{Hz}$$

where

E = plate material modulus of elasticity, psi

t = plate thickness, in.

w = material weight per cubic foot, lb/ft^3

a = length, in.

b = width, in.

If the ratio of the panel length to width is c = a/b, the expression in parentheses can be simplified by setting a = Cb:

$$\frac{C^2 + 1}{C^2 b^2} = \frac{F}{b^2}$$

The factor F has a maximum of two for a square plate and decreases as the ratio c increases. It is 1.1 when the length is three times the width.

The (1,1) resonant frequency then becomes

$$f_{1,1} = 368 \frac{tF}{b^2} \sqrt{\frac{E}{w}}$$

The ratio t/b^2 that makes this equation equal to 500/3 is the maximum which ensures that the design curve is valid at 500 Hz. This ratio is a very small fraction, so it is more convenient to deal with its inverse. This inverse ratio was determined for a number of materials, for four length-to-width ratios, and are listed in Table 9.2. The tabulated values are minimums that should be considered in sizing enclosure panels of any selected thickness.

9.3.4 Absorption Materials

All materials absorb some sound. A smooth palstered wall, for example, has an absorption coefficient of around 0.02 and plywood has coefficients above 0.10. Materials must be porous to be good sound absorbers, however. Sound pressure waves can then move air molecules back and forth across solid surfaces and convert their kinetic energy into heat. It might be well to point out that since sounds generally have only a fraction of a watt of power, the heat generated does not cause problems.

The resistance to airflow provided by a material is an indication of its absorptive ability. If a material has low resistance, sound waves passing through it are not attenuated very much. In contrast, if the resistance is very high, very little of the sound penetrates and, again, very little attenuation occurs. Flow resistance is a function of fiber diameter, density, structure, and the binder used to hold it together.

The sound absorption coefficeint of a material is the fraction of energy absorbed when the sound passes through it, is reflected, and passes through again. Perhaps the term "absorptive system" should be used here because in this context, the nearby reflecting surface is considered to be a part of the material. Similarly, space between the reflector and absorber as well as the means holding the

TABLE 9.2 Minimum Enclosure Panel Size

	b^2/t			
Material	c = 1	1.5	2	3
Aluminum	1200	840	730	640
Chipboard	440	320	280	240
Fiberglass	570	410	360	320
Glass	1200	850	740	650
Gypsum board	1500	1100	960	840
Hardboard	540	390	340	300
Lead	270	200	170	150
Plexiglass	400	300	250	220
Plywood	660	470	400	360
Steel	1200	820	720	620

two elements together are also taken to be part of the system. Absorption material systems sometimes also include resonant cavities that reduce sound by cancellation.

Figure 9.16 shows how the absorption of a typical material varies with frequency. This material is fairly porous, so that at frequencies above 1000 Hz, sound waves freely move air molecules in and out of the pores, causing them to rub against internal surfaces and convert kinetic energy to heat. At lower frequencies, the material itself moves in response to the sound waves, so less rubbing and energy conversion occurs. As this occurs, absorption roughly decreases with frequency. This fall-off starts at a frequency whose wavelength is about 16 times the material thickness.

Normally, with hydraulic noise, there is little interest in attentuating low frequencies. In special cases where frequencies at pump shaft frequencies must be reduced, the low effectiveness of absorptive materials is a problem. The obvious course, suggested by Figure 9.16, is to increase the thickness of the material. It can be seen that this has the effect of moving the curve to the left. Near-maximum absorption for a material is provided when the material thickness is increased to one-sixteenth of the wavelength at that frequency. This thickness, in inches, is approximately

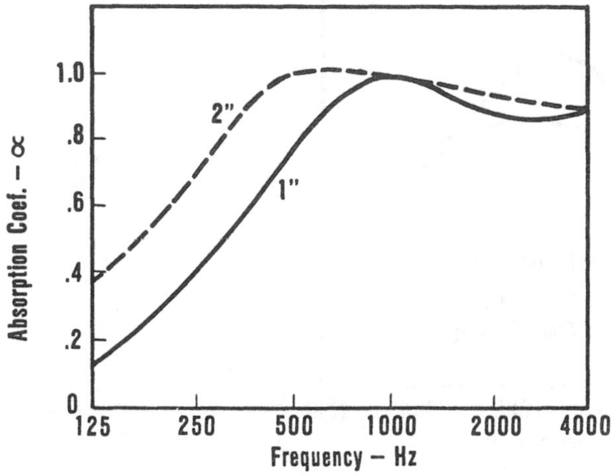

FIGURE 9.16 Typical material absorption curves. (From Ref. 4.)

$$t = \frac{850}{f} \quad \text{in.}$$

where f is the frequency where high loss is required in hertz.

Actually, the critical factor is not the absorptive material thickness but is the distance from the reflector to the outer face of the absorber. Figure 9.17 shows the effect of moving an acoustic tile 6 in. from the reflector. This has the effect of shifting the drop-off frequency to the vicinity of 125 Hz, where it would be if 6 in. of material had been added. It can be seen that although absorption remains high down to this frequency, it is somewhat lower than the original maximum. The poor absorption at high frequencies is due to insufficient porosity on the face of the tile.

Sound absorption coefficeints for architectural materials are available from many sources, but these do not provide data for many of the materials that are preferred for machine noise control. A compendium of noise control materials (5), published in 1975, is a good source of information on products available at that time. The best sources for data on newer materials are the manufacturers of these products. A listing of acoustic material suppliers given each year in the July issue of *Sound and Vibration* magazine (6) is useful in contacting these sources.

Published absorption coefficients usually include values for 125, 250, 500, 1000, 2000, and 4000 Hz. Sometimes only the NRC rating, which is the average of the coefficeints at the four frequencies

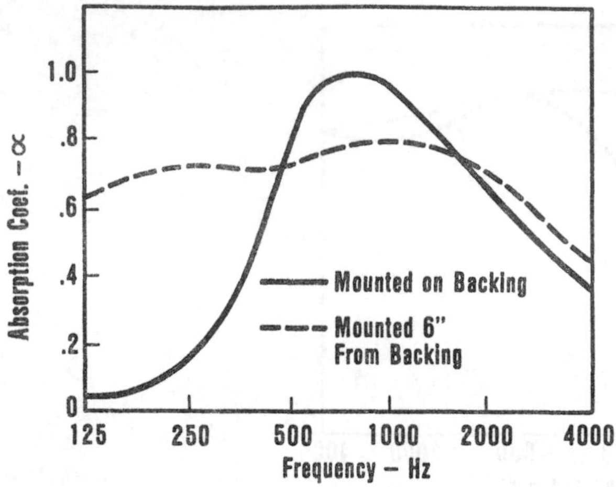

FIGURE 9.17 Effect of air space behind acoustic tile. (From Ref. 4.)

between 250 and 2000 Hz, is given. This average value is conservative for hydraulics since it includes the data for 250 Hz, which is not relevent and is usually quite low.

Material thickness, mounting details, and spacing from the reflecting surface all affect absorption, so care must be taken to note these factors in reviewing material ratings.

There are many types of sound absorption materials on the market. Some must be ruled out for use in hydraulically operated machines because they are a fire hazard. Although they may not even support combustion under ordinary conditions, they act as wicks when splashed with oil. It should also be noted that materials saturated with fluids lose much of their sound absorption ability. Cotton, paper, wood, wool, hair, and plaster products should be avoided for these reasons.

All porous materials tend to store fluids. Some, like fiber glass wool and metal felt, will eventually drain and lose their flammability. It is not known which category mineral wool is in. Plastic sponge materials appear to have a variety of structures, so they may be of both types. It is advisable to ask manufacturers about the oil retention characteristics of their products or to conduct in-house tests before these materials are used.

Membranes are used to give absorptive materials protection from fluids. Some materials, particularly polyurethane foams, are sold with such protection. The disadvantage of such films is that their mass provides a barrier to sound which reduces absorption at frequencies above 1000 Hz.

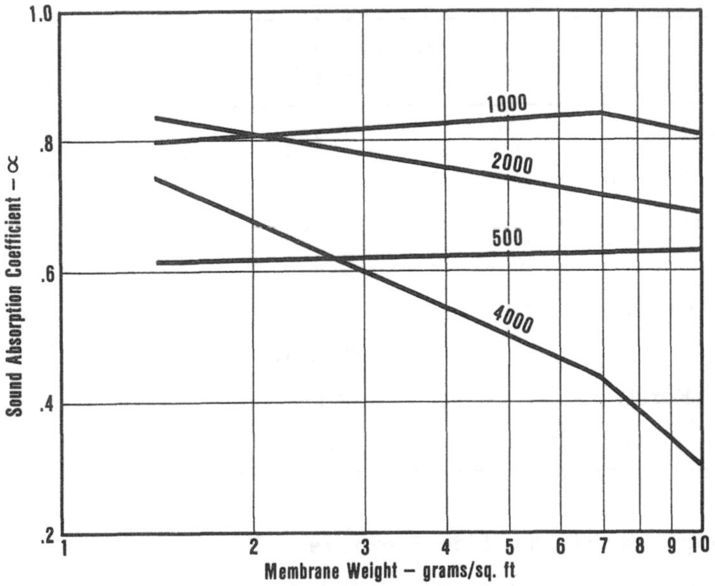

FIGURE 9.18 Effect of membrane mass on absorption. (From Ref. 7.)

Figure 9.18 shows how the mass of the film modifies the performance of an absorption material (7). Although presence of a film decreases effectiveness at higher frequencies, it enhances it at lower ones. As the chart shows, the absorption at 500 Hz is increased as film mass is increased. At lower frequencies this effect is even greater.

Table 9.3 gives the weights of various film materials. With reference to Figure 9.18, it can be seen that 0.001-in. films of most of these materials will have little effect. In most cases, even 0.0015-in. films would be satisfactory, and for the lighter materials 0.002-in. films probably can be used.

Absorptive materials are generally fragile and require mechanical protection. They can be covered with perforated sheet metal or Masonite, which have holes accounting for at least 20% of the surface area, without reducing effectiveness, even at high frequencies. Expanded metal, which is very open, is also used frequently.

The holes in a facing, in combination with the air in the absorption material, act as a Helmholz resonator (8,9) that greatly reduces sound at the resonant frequencies. This is usually effective at very low frequencies, so it is seldom useful in quieting hydraulics.

TABLE 9.3 Film Materials

Material	Weight/0.001 in. (g/ft^2)
Polypropylene	2.15
Polyurethane (Tuftane)	2.36
Polyvinyl fluoride (Tedlar)	2.56
Polyethylene	2.84
Aluminized polyester	2.88
Polyester	2.97–3.47
Polyurethane (Korel)	2.98
Plasticized polyvinyl chloride	3.41

9.3.5 Cladding Materials and Line Enclosure

Cladding is a flexible barrier material that has a layer of polyurethane foam or fiberglass bonded to it. This material is often applied directly to hydraulic lines, pumps, and other noise sources to enclose their noise emissions.

The most common cladding material is filled vinyl. It is made by mixing powdered high-density materials with vinyl and rolling it into sheets. It is believed that lead-filled vinyl was developed first to shield X-ray technicians. Later, when it was found to be useful as a sound barrier, lower-cost fillers were introduced. Sheet lead and polyvinyl films are also used in cladding.

Weights of the barrier layers in these materials range from 0.33 to 3.0 lb per square foot. One pound per square foot appears to be the most popular.

The foam or fiberglass layer is generally 0.25, 0.5, or 1.0 in. thick. It is installed in contact with the noise source and mechanically isolates the barrier from the source. As with conventional enclosures, direct contact transfers acoustic energy to the barrier.

The greatest difficulty encountered in applying cladding is in sealing joints. Two types of joints are used. One is a butt joint, as shown in Figure 9.19. When this method of joining is used the barrier layer must be carefully fitted without gaps. Since voids in the resilient layer have no effect, it is advisable to bevel it and make it easier to avoid holes in the outer layer. When lead tape is used to join the parts, it ameliorates the effects of fitting imperfections.

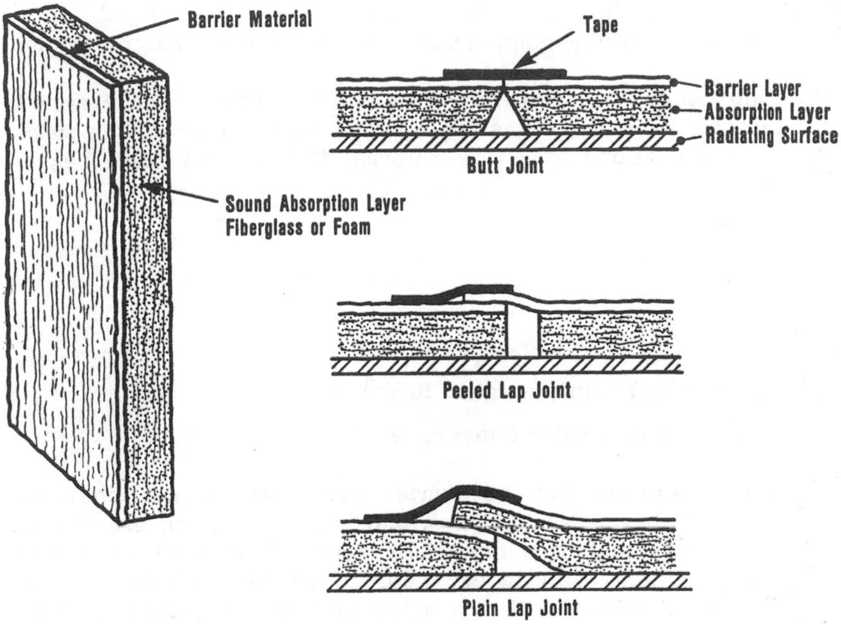

FIGURE 9.19 Cladding material and joints.

The lap joints shown in Figure 9.19 are also used. In making lap joints it is best to remove some of the inner layer so that only the barrier layers overlap. This is difficult to do and installers try to avoid this step. Tight joints are possible without this removal, but they are bulky and require extreme care. A novice using this type of joint in cladding lines for a pump test, for example, may cause the test levels to be 4 to 5 dB higher than they would be if all the cladding joints are tight. Some laboratories avoid trouble by using two layers of cladding on test lines. Any type of tape can be used for lap joints, although lead tape probably is best.

Cladding materials come in flat sheets. Since it is difficult to make good joints with it, when enclosing irregular shapes, its use is generally limited to very simple shapes.

Cladding is probably most commonly used for enclosing hydraulic lines. The value of this practice is seen in results of pump noise evaluation tests. Test codes requires that only the noise radiated by the pump itself be measured. When the test lines are properly cladded, the measured noise levels sometimes drop more than 6 dB, indicating that the lines are radiating more noise than the pump.

In pump testing the pump itself is often enclosed in cladding to check for extraneous noise. If sound levels do not increase at least

6 dB when the cladding is removed, the test space is judged as having too high a background noise level for testing that particular pump.

Cladding behaves as a spring-mass system; with the air trapped within the soft layer acting as the spring and the barrier the mass. When a line is wrapped with this material, it has a resonant frequency of

$$f_n = \frac{120}{\sqrt{wt}} \quad \text{Hz}$$

where

w = barrier layer unit weight, lb/ft^2

t = thickness of resilient layer, in.

This equation assumes that the barrier layer does not contribute to the stiffness of the cladding system and the only spring effect comes from the air entrapped in the resilient layer. Care must be exercised in installing cladding to avoid crushing the resilient material, as this decreases the air volume and raises the resonant frequency. Crushing also increases the energy transfer to the barrier.

Below about 1.5 times the resonant frequency, the cladding provides no noise attenuation. Above this frequency it operates in the mass-controlled region and its transmission loss is as shown in Figure 9.15. This curve is for material weighing 1 lb/ft^2. To use it for materials of other weights, multiply the frequency scale by their unit weights.

As with enclosures, the noise reduction afforded by lagging, the insertion loss, is the transmission loss less the reverberant sound pressure that builds up between the source and the barrier. The resilient layer absorbs sound and reduces this reverberation. However, with the thicknesses provided on most lagging, absorption is poor at low frequencies. The reverberant noise level is almost as high as the transmission loss at these frequencies, so there is little noise reduction.

Absorption increases with frequency until reverberation is not significant and the noise reduction equals the full mass transmission loss of the barrier. This occurs at about 1000 Hz for a 1-in. absorptive layer and 2000 Hz for a 1/2-in. layer.

From 1.5 times the resonant frequency, where no noise reduction occurs, to the critical frequency, the noise reduction in decibels rises linearly with frequency. Above the critical, noise reduction is the same as the transmission loss given in Figure 9.15.

A very neat hydraulic-line enclosure system (10,11) has been developed by BHRA Engineering in England. It uses polyvinyl

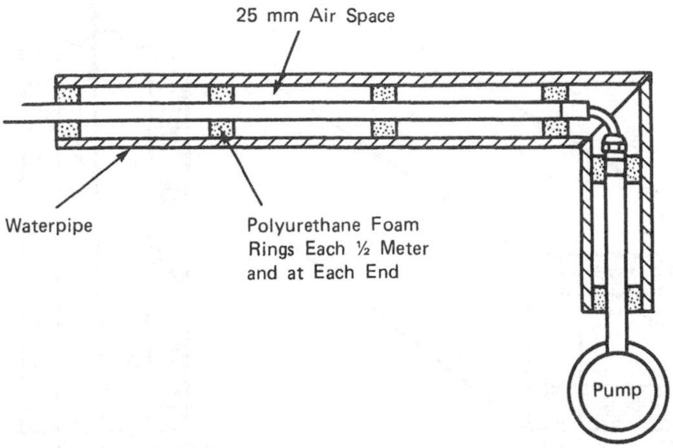

FIGURE 9.20 BHRA hydraulic-line enclosure system. (From Ref. 10; with permission of BHRA, The Fluid Engineering Centre.)

chloride (PVC) water pipe as shown in Figure 9.20. This material, in 3- and 4-in. sizes, has a weight of around 2 lb/ft^2.

The line is kept centered by polyurethane foam rings spaced 1/2 meter apart. Concern is expressed that more rings may transmit more energy from the line to the pipe, but it would appear that noise reduction might be enhanced if more absorptive material was put into the pipe. Pipe ends are closed with the foam rings. They find that most of the sound is radiated from the first 1 or 2 m of pump and valve discharge lines, so the enclosures are only installed on these limited lengths.

The pipe can be slit lengthwise for installation and held together with tape. Bends are made with PVC fittings or by mitering. Since the pipe is free of vibration, it can safely be attached to other static machine structure.

Both lagging and the BHRA line enclosure system are very difficult to apply. There have been commercial lagging systems offered in the past that were easier to apply and made a good appearance. It is not known how effective they were. For some reason they have not gained acceptance and appear to have dropped out of the marketplace. The benefits of line enclosure has been well demonstrated and it is a mystery why commercially available, easily applied laggings are not commonly used.

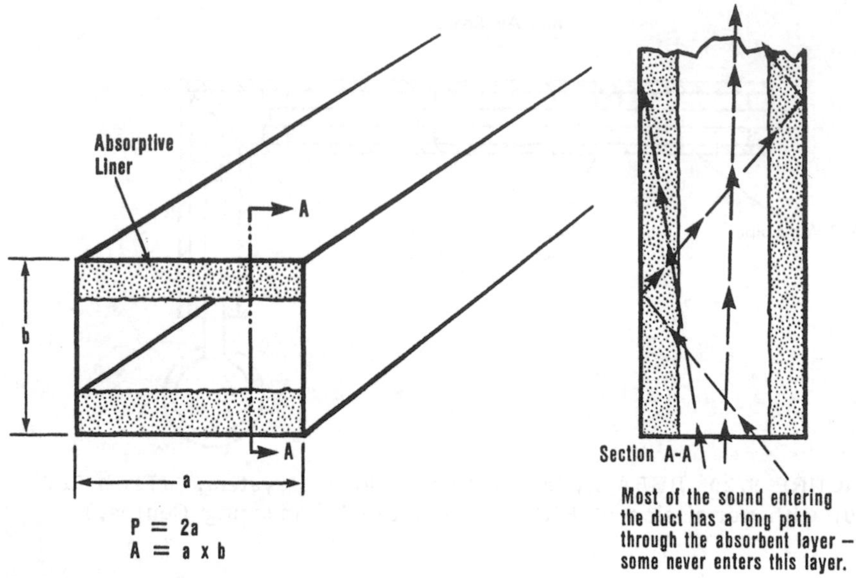

FIGURE 9.21 Sound absorption in air ducts.

9.3.6 Air Ducts

Enclosures normally depend on almost watertight sealing for their effectiveness. As discussed earlier, even small cracks along parting lines and around hydraulic lines reduce their noise reduction capability. This approach obviously cannot work where large openings must be provided for cooling air. In such cases sound exiting through the large openings must be attenuated before reaching the environment outside the enclosure.

Muffling is usually accomplished with ducts lined with absorptive material, like that shown in Figure 9.21. Sound enters these ducts at all angles. Some, at oblique angles to the walls, is readily absorbed, as it has to travel through the length of the acoustic lining. Sound at greater angles will have similar long paths through the liner, as it reflects back and forth between a pair of walls. Sound parallel to the walls that does not pass through the lining is never absorbed.

The attenuation in a lined duct is calculated from (12)

$$AT = 12.6 \frac{P}{A} \alpha^{1.4} \quad \text{dB/ft}$$

where

P = duct-lining perimeter, in.

A = cross-sectional area, $in.^2$

α = linear absorption coefficient

If the duct has absorptive material on all its walls, the factor P is the duct perimeter. If the duct is only partially lined, P is the total of the wall widths or heights that are covered. Splitters, absorptive material baffles parallel to the airflow, are sometimes added to ducts to increase absorption. This increases P by twice their width, because each side of the baffle acts like material on a duct wall. For the same reason, the splitter material acts as though it is one-half its actual thickness. Since thickness affects the absorption coefficeints at lower frequencies, it is advisable to make splitters twice as thick as the wall lining to ensure that all the absorption material has the same frequency characteristics.

Absorption coefficients for a given material depend on how the material is mated with its reflective surface. For this equation the values that must be used are for material mounted on the reflective surface on 1 × 3 firring strips. In tables of architectural acoustic materials, this is referred to as "AIMA No. 6 mounting."

A fairly large percentage of the noise entering a duct is parallel to the ducts length and is not absorbed because it never enters the lining. For this reason the duct should have at least one lined right-angle bend in it. If the bend follows a straight section, only parallel sound reaches it. All the sound, then, strikes the wall at right angles, so it is partly absorbed and partly reflected backward. Some diffraction occurs in this process, so it is advisable to have another straight section after the bend. The straight sections should have lengths equal to or greater than three times the duct width.

There is a large selection of commercially available mufflers. These provide high attenuation over a wide band of frequencies and have reasonable airflow resistance. They are expensive but they reduce the engineering effort needed to develop a quiet machine. They are also relatively compact, so they shorten air ducts and minimize machine size. It is recommended that their use be explored whenever duct silencing is required.

REFERENCES

1. A. Crook and R. A. Heron, "Airborne Noise from Hydraulic Lines Due to Liquid Borne Noise," paper C263/77, *Proceedings of the Quiet Oil Hydraulic Systems Seminar*, Institution of Mechanical Engineers, Nov. 1977, pp. 81–91.

2. M. L. Hughes, "Flexural Vibrations in Pipework Due to Liquid Borne Noise," paper C260/77, *Proceedings of the Quiet Oil Hydraulic Systems Seminar,* Institution of Mechanical Engineers, Nov. 1977, pp. 51–58.

3. *Improved Sound Barriers Employing Lead,* A. I. A. No. 39, Lead Industries Association, New York.

4. R. F. Lambert, "Sound in Large Enclosures," in *Noise Reduction,* L. L. Beranek, ed., Mc-Graw-Hill, New York, 1960, Chap. 11, pp. 222–245.

5. *Compendium of Materials for Noise Control,* HEW publication (NIOSH) 75-165, U.S. Department of Health, Eduction, and Welfare, U.S. Government Printing Office, Washington, D.C., June 1975.

6. *Sound and Vibration,* Bay Village, Ohio.

7. R. K. Miller and W. V. Montone, *Acoustical Enclosures and Barriers,* Fairmont Press, Atlanta, GA, 1978, p. 23–26; also W. R. Powers and C. D. Rudinoff, "The Noise Box Test," NOISEXPO, 1975.

8. D. B. Callaway and L. G. Ramer, "The Use of Perforated Facings in Designing Low Frequency Resonant Absorbers," *Journal of the Acoustical Society of America* (3): May 1952.

9. E. E. Mikeska and R. N. Lane, "Measured Absorption Characteristics of Resonant Absorbers Employing Perforated Panel Facings," *Journal of the Acoustical Society of America* 28(5): Sept. 1956.

10. R. Heron, "Acoustic Enclosure and Cladding," in *Quieter Fluid Power Handbook,* BHRA, The Fluid Engineering Centre, Bedford, England, 1980, Chap. 13, pp. 95–99.

11. BS 5944, *Measurement of Airborne Noise from Hydraulic Fluid Power Systems and Components: Part 3. Guide to the Application of Part 1 and Part 2,* Section 6.5 Acoustic Cladding, British Standards Institution, London, 1980.

12. R. W. Leonard, "Heating and Ventilating System Noise," in *Handbook of Noise Control,* C. M. Harris, ed., McGraw-Hill, New York, 1957, Chap. 27, pp. 27–29.

13. *More Sound Advice,* Vickers Inc., Troy, Mich.

14. *Quiet Please,* Vickers Inc., Troy, Mich.

10
Mechanical Isolation

Mechanical isolation is one of our most effective hydraulic noise controls. As pointed out in Chapter 7, pumps and motors generate very high levels of structureborne noise. If even a small percentage of this energy reaches the radiation sources discussed in Chapter 9, these radiate higher sound levels than those coming directly from the pump or motor. Mechanical isolation is the technique used to interrupt the structureborne noise energy flowing from the source to a radiator.

10.1 THE ISOLATION PRINCIPLE

The principle of vibration isolation is very simple. In Section 3.1.3 we saw that if a sinusoidially varying force is applied to a mass supported on a spring, as shown in Figure 3.1, the displacement of the mass is

$$x = \frac{(DF)p \sin \omega t}{k} \quad \text{in.}$$

where

DF = dynamic factor governed by the ratio of the forcing to natural frequencies

K = spring rate, lb/in.

p = force amplitude, lb

ω = force angular frequency, Hz

This is the same equation that was used in discussing structureborne noise in Section 7.2.1, except that we chose to use English instead of SI units here.

The force transmitted to the surface supporting the spring is equal to the displacement times the spring rate. The *transmissibility* of the spring-mass system is the ratio of this force to the applied force

$$T = \frac{xK}{p \sin \omega t}$$

Substituting the earlier expression for displacement x, we see that the transmissibility is equal to the dynamic factor

$$T = DF$$

As seen in Figure 3.3, at low frequencies the dynamic factor, hence the transmissibility, is 1. All the force applied to the mass reaches the surface under the spring. At higher frequencies, above 1.4 times the natural frequency of the spring-mass system, the mass opposes the applied force and transmissibility decreases with increasing frequency.

The transmissibility curve in Figure 10.1 is a new plot of the zero-damped curve of Figure 3.3. The scale used shows more clearly how effective isolation can be. Isolation mounts have moderate damping, which has little effect on the isolation part of the curve. Although this justifies using the zero-damping curve, it should be kept in mind that all the curves in the earlier figure depict transmissibility.

Vibration isolation is then the practice of mounting masses on resilient members, which gives them natural frequencies well below the frequencies of the loads acting on them, so that very little of the forces reach the structures supporting the masses.

10.2 ISOLATION SYSTEM DESIGN

Since pumps, motors, and valves are structureborne noise sources, they are also candidates for vibration isolation. Valves are the easiest to isolate. They generate high-frequency noise, and as we will see, this is the easiest to isolate. More important, their isolation is not made complicated by drive-shaft considerations, like that of pumps and motors.

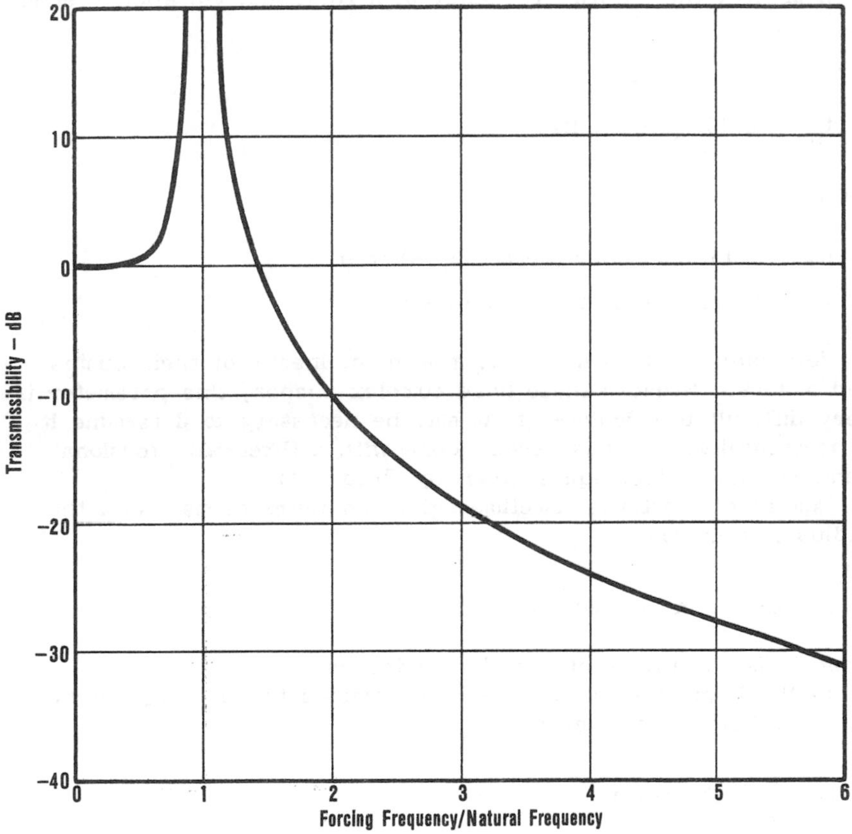

FIGURE 10.1 Transmissibility of vibration isolators.

10.2.1 Natural Frequency

In Section 3.1.2 we saw that the natural frequency of a spring-mass system is

$$f_n = 0.159 \sqrt{\frac{K}{M}} \quad \text{Hz}$$

where

K = spring rate, lb/in.

M = mass, lb-sec^2/in.

Similarly, the natural frequency of a rotational vibration system is

$$f_{nr} = 0.159 \sqrt{\frac{K_r}{I}} \quad \text{Hz}$$

where

K_r = rotational spring rate, in.-lb/rad

I = moment of inertia, in.-lb-sec^2

Few manufacturers know the moment of inertia of their pumps and motors. Because these have complex shapes, this parameter is very difficult to calculate; so it may be necessary to determine it experimentally. This is usually done with a three-wire torsional pendulum using techniques given by Crede (1).

Moment of inertia is sometimes given in terms of mass and the radius of gyration

$$I = Mr^2 \quad \text{in.-lb-sec}^2$$

where r is the radius of gyration in inches.

In the English system, mass is determined by dividing weight by the acceleration of gravity

$$M = \frac{w}{g} \quad \text{lb-sec/in.}$$

where

w = weight, lb

g = acceleration of gravity 386 in./sec^2

So natural frequencies can be found from

$$f_n = 3.13 \sqrt{\frac{K}{w}} \quad \text{Hz}$$

$$f_{nr} = 3.13 \sqrt{\frac{K_R}{wr^2}} \quad \text{Hz}$$

For linear vibration, another form of the equation is also useful. It is determined by considering the mass resting on its resilient mount, the amount that it compresses the isolator is referred to as the *static deflection*. This is equal to

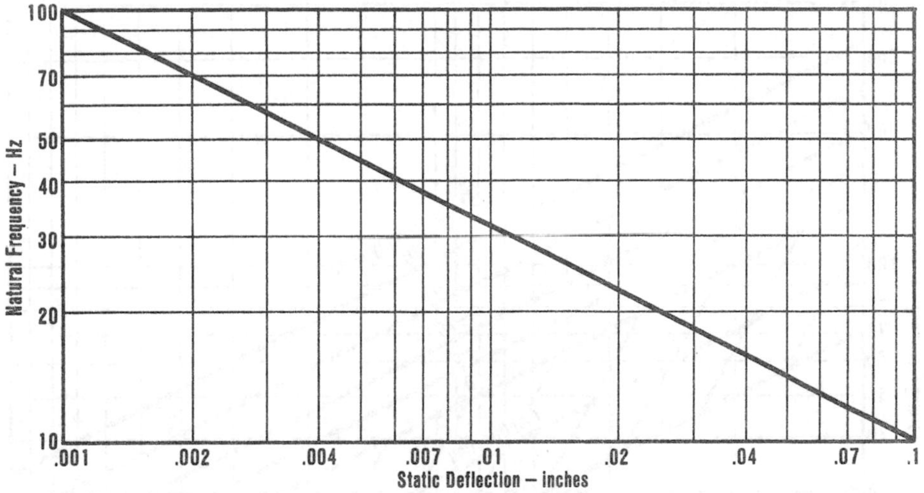

FIGURE 10.2 Relation between natural frequency and static deflection.

$$\delta_{st} = \frac{W}{K} \quad \text{in.}$$

Natural frequency can then be expressed in terms of this quantity

$$f_n = 3.13 \sqrt{\frac{1}{\delta_{st}}} \quad \text{Hz}$$

This relationship is shown in Figure 10.2. The effect of static deflection on transmissibility is shown in Figure 10.3.

These equations are based on the assumption that the support for the spring-mass system is infinitely stiff. This, of course, cannot be achieved in practical machines. However, most of the predicted isolation will be achieved if the lowest natural frequency of the support is more than five times that of the isolators.

In cases where one of the strong isolated frequencies is capable of exciting resonance in the support, more attenuation may be required to make a machine quiet. This can be achieved by increasing the support stiffness to detune it. If this is not practical, the mount stiffness should be decreased to where the natural frequencies ratio is 10 to 20.

In Chapter 8 valve noise was said to exist at all frequencies but is greatest in the range 4 to 8 kHz. Valve mountings with natural

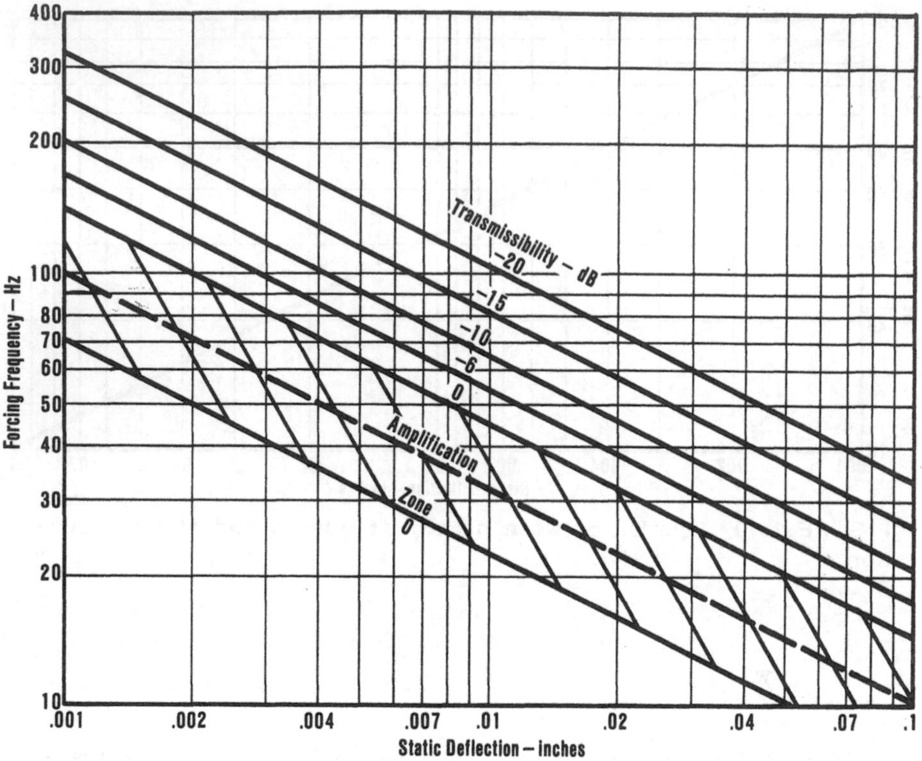

FIGURE 10.3 Transmissibility as a function of static deflection.

frequencies of several hundred hertz, then, very effectively isolate this noise. The static deflection for such mounts is only a few thousanths of an inch, which is easily provided.

Selecting the natural frequencies for pump and motor mountings is not as easy. In Chapter 7 it was said that the structureborne noise spectra of these units are very similar to the fluidborne noise spectrum in Figure 6.6, so we will use this figure for the basis of this discussion. Because of the frequency weighting factors involved in airborne noise, the second and third piston harmonic components of this spectrum are the most likely to excite other machine components into radiating loud noise. The spectrum also has a very high peak at piston frequency and significant peaks at the shaft fundamental and its second and fourth harmonics.

While the objective of mounting this pump is to isolate the two critical noise components, care must be exercised to prevent enhancing the strong lower-frequency components. The pumping frequency, if amplified enough to overcome the frequency weighting effects, could become a noise problem. Similarly, the shaft frequency and its harmonics could cause uncomfortable vibrations and even structural fatigue failures.

Resonance amplifies forces with frequencies from about 0.7 to 1.4 times the natural frequency of an isolation system. The highest of these frequencies is twice the lowest, so the amplification zone for any mounting system is one octave wide. It should be noted that the natural frequency of the system is not at the center of this zone but is at 1.4 times the lower limit and 0.7 times the upper limit.

If we refer again to Figure 6.6, a number of general observations can be made. There are a number of windows, at least an octave wide, where it would be safe to establish isolator natural frequencies. The highest is between the first and second pumping harmonics. Mounts tuned to the appropriate frequency in this zone will not attenuate either of these two harmonics but will reduce the third by 5 dB. The advantage of this window is that it only requires a little resilience between a pump and its support. This is often easy to provide. It is not suitable, however, where the first two pumping frequencies are strong or are in the critical range.

The next-lower window occurs below the pumping frequency. As seen in Figure 6.6, the lower limit is the shaft fourth harmonic, so this window exists only for pumps or motors with eight or more pumping chambers. Usually, the shaft fourth harmonic is not as strong as shown in Figure 6.6. In these cases the window extends down to the second shaft harmonic and therefore exists for pumps and motors with four or more chambers.

If the mounting system is tuned to 0.7 of the pumping fundamental, it does not attenuate this frequency but does reduce the second pumping harmonic by about 9 dB. For shaft speeds of 3000 rev/min and above, the pumping frequency is in the critical range and its attenuation may be necessary. Where the window is more than one octave wide, this can be provided by lowering the natural frequency. This also increases the reductions of the harmonics.

Static deflections in this window are in the range 0.02 to 0.1 in. These are easily attained with commercially available mounts. They are also attainable with simple resilient pads and with "homemade" mounts.

The next lower window occurs between the shaft frequency and its second harmonic. Mounts designed for this zone give very good attenuation to all pumping harmonics but none or little to the shaft harmonics. Its use appears to be most appropriate for very high-speed units if they do not have strong shaft harmonics.

The lowest-frequency window is below the shaft frequency. With industrial equipment pumps this is below 20 to 30 Hz. Screw pumps are an exception having shaft frequencies of 60 Hz. Mobile pumps generally operate at higher speeds, so their shaft frequencies usually range from 30 to 50 Hz. Aircraft pumps range from 40 to over 100 Hz. Motor speeds vary over the range from a few to 100 Hz.

The advantage of this zone is that it provides some attenuation for the shaft harmonics as well as high reductions in the pumping harmonics. It is well suited for high-speed units. These require only nominal static deflections. Lower-speed units must have much higher deflections that are sometimes difficult to achieve with standard mounts. Very soft mounts tend to be unstable, and provisions for locking them out must be provided for times when the machine is shipped or even moved. They also deflect excessively with torque changes, and this can cause misalignment problems.

Very soft mounts on mobile machines require additional consideration. Shocks from running over rough surfaces produce deflections that cause misalignment and clearance problems. A 1*g* acceleration generated by going over a bump will cause a vibration amplitude equal to the static deflection. Fortunately, most transportation shocks are much less than this level. Vibration caused by travel over rough ground or by tire lugs can also excite mount resonances. Generally, these excitations have relatively low frequencies that seldom reach 10 Hz.

10.2.2 Short Circuiting

If a pump or motor is resiliently mounted but is connected to the machine circuit through steel lines, it is not truly isolated. It is not free to vibrate as assumed in the earlier analyses, so its mass cannot react against the applied force as predicted.

There are two ways to look at this situation. One is that the mounting is effectively stiffer because of the line resistance. The other is that while the mounts provide isolation, the tubing provides a parallel transmission path for the vibration. The latter concept gives rise to the term *short circuiting,* which is applied to such cases. Both concepts are simplifications that are not strictly correct but are handy in analyzing machine situations.

Any time a hydraulic component is isolation mounted, all connections between it and stationary points must be more flexible than the mounts. Flexible hydraulic hose must be substituted for solid lines. The configuration of such lines is important and is discussed in more detail later in this chapter.

Similarly, flexible cable must be used instead of metal electrical conduit. The only rule for using this material is that it must permit free motion in all directions.

Mechanical links sometimes used to control pump displacement also cause short circuiting. The effectiveness of an isolator in reducing transmitted vibration through a link of this type is discussed in Section 10.5. That discussion pertains to a solidly mounted pump. If the pump were isolation mounted, a more flexible isolator would have to be used. This added flexibility would make it impossible to control precisely the speed of the farm machine in which the manual control is used. In some cases the flexibility of a pump mounting and the control stiffness can both be preserved by using a fairly stiff control isolator, but locating it farther from the pump so that its deflections are amplified by the resulting long lever arm.

Flexible control cable or Boden wire can also be used to control pumps without short circuiting isolators. These devices consist of a wire that moves axially within a spirally wound conduit that is anchored at both ends. They have excellent transverse flexibility but are stiff axially. When they are looped for at least a quarter turn, they have good flexibility in all directions.

10.2.3 Isolators

There are a great number of different types of commercial vibration mounts. There is no one best type. Generally, there are so many meeting a given set of basic requirements, selection is based on cost and how well the mount fits the rest of the machine's design.

Mounts are not standardized, so manufacturers' catalogs must be consulted in making a selection. Many suppliers also provide free consultation when requested. Contacting these sources is made easy by design and engineering magazines which carry the manufacturers' advertisements. Each year *Sound and Vibration* magazine (2) publishes a list of sources in their July issue. The list is comprehensive and there is difficulty in winnowing out the ones offering suitable products.

Isolators tend to be expensive. Part of their cost is in the technical services provided by manufacturers. Some is due to manufacturers offering such a wide variety of mounts that it is not possible for them to achieve the economies that come from making great quantities of identical units.

There are manufacturers making mounts for high-volume products, such as automobiles, that have much lower production cost. Their prices often provide a great temptation for using their products. However, these sources only make the mounts. They do not engineer them and cannot provide adequate design data for their use. So these products must be evaluated by trial. This, of course, increases the design and development cost of a machine. Since there is usually only one design and stiffness offered, tuning is not possible. Use of such mounts also has the liability of their becoming unavailable when their original high-volume customer quits buying them.

Most mounts are made of natural or synthetic rubber. These materials are molded into many different shapes to carry loads by compression, shear, tension, or bending. This provides the needed wide diversity of characteristics. Further, their compounding is varied to change stiffness so that mounts with a given size and shape are made with a range of spring rates.

Rubbers have appreciable hysteresis, so they provide damping. In some applications this is a desirable feature because it keeps vibration amplitudes within reasonable limits during startup and shutdown, when the excitation passes through the mount resonant frequency. This generally is not too important in hydraulic machines because excitations are usually not high at these times.

Rubber mounts do not have constant spring rates. Generally, they are designed to have a nearly uniform rate in the vicinity of their rated load, and nonlinearity is no problem. In this load range, their static deflections should be taken as the load divided by this rate rather than their actual static deflection. At low loads their rates are lower than at rated loads, and it may take some digging to find the correct rate at these loads. Overloading mounts greatly shortens their life, so it should never be done. For this reason the higher rates at higher loads are not a concern.

Natural rubber and some synthetics deteriorate when exposed to petroleum products. For this reason Buna N and Neoprene rubbers, which have good resistance to oil, are preferred for hydraulic machines.

Metal springs are also used as isolators. They withstand greater temperature extremes than do rubber springs. More important from a hydraulic machinery standpoint, they provide much greater deflections, thus lower natural frequencies. In some cases these low natural frequencies can be secured with rubber mounts by mounting one on another so that their static deflections add.

Metal springs are very linear and the classical isolation equations apply to them without corrections. They also have very little hysteresis, so mounts often contain friction devices to control vibrations while going through resonance.

Spring mounts must also have a layer of resilient medium, such as rubber or cork, if they are used for noise isolation. At high frequencies metal springs are good transmitters of some types of structureborne acoustic waves (3). This type of transmission short circuits the isolation mechanism that we have been discussing. Interposing the soft medium into the load path provides a very large change in acoustical impedance. As discussed in Chapter 6, such impedance changes reflect the acoustic waves back toward their source.

Isolators can also be made from sheets of resilient materials. Materials made for this purpose include sheets of Neoprene molded into wafflelike grids, sheets and tubes of elastomer-impregnated canvas, sheets of sandwich consisting of cork faced with an elastomer, and felt cloth.

The simplest way to use such material is to cut it into pads and place it under a component's feet or mounting pads. No fasteners are used and gravity is counted on to keep the component in place. The material usually has some sort of nonskid surface, so this scheme works well for isolating motor-pump assemblies and whole machines.

There is a tendency to make the pads as large or larger than the machine surfaces bearing on them. Usually, these surfaces have large areas that spread the load out to where the pads have only low unit loads. This practice does not utilize the full potential of the material. Pad size should be determined from the load that must be carried and the material load rating. This provides the rated static deflection for the material. If a greater deflection is required, a number of thicknesses of the material are used to secure it.

When the support surfaces of a component are small in relation to the load capacity of the isolation material, metal plates are used to spread the load over a greater area. Some manufacturers offer their material bonded to metal pads that can be used in such cases. Usually, these are also equipped with a threaded stud for bolting them directly to the isolated unit.

The use of these materials is not restricted to applications where gravity keeps the isolated unit in place. Figure 10.4 shows a mounting, made of these materials, that works well as a pump support. Impregnated canvas was used, but with modification other materials could also be employed.

The impregnated canvas was used because it also comes in tube form which is useful in preventing metal-to-metal contact between the bolt and the pump foot. Such contact would be a short-circuit path for the acoustic waves that were discussed in connection with metal spring isolators.

In these mounts, resilient material must also be placed on the top as well as the bottom of the pump foot to prevent metal-to-metal contact. The bolt preload causes the two resilient members to act in parallel, so the mount stiffness is the sum of their two stiffnesses.

Care must be taken to keep the bolt from being tightened too tightly. This overloads the resilient material and reduces its life. Also, since this material does not have a constant spring rate, increasing its loading makes it stiffer and raises the isolator's natural frequency.

Since the bolt is not fully tightened, it cannot develop the friction forces that normally keep it from vibrating loose. It can be argued that the bolt will not loosen because it is not heavily loaded. Regardless, it is felt that it is good practice to use a locking method for keeping it from backing out.

These mounts are very effective when properly designed. They are particularly useful in troubleshooting and development work in

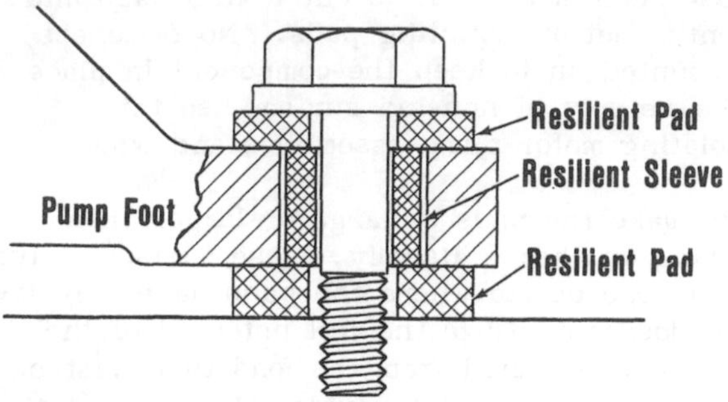

FIGURE 10.4 Fabricated isolator.

the laboratory because they can be produced faster than commercial mounts can be secured. They are not as cheap as they look, however, and their actual cost should be compared to that of equivalent commercial mounts before it is decided to use them in production.

10.3 PUMP ISOLATORS

This section focuses on isolating pumps. It applies equally well to motors that have most of the same problems.

Previous discussions treat isolators as though they are like the single spring in Figure 3.1. For obvious reasons they must consist of a number of units distributed about the isolated mass. Their locations and individual characteristics are dictated by such factors as stability, shaft torque, and isolation goals.

This discussion of pump isolators follows two tracks. One relates to the American practice of mounting a pump and its drive motor on a common plate. This part of the discussion also pertains to isolating a whole hydraulic power supply from the rest of a machine and to isolating entire machines from building floors or ship decks.

The second track relates to the British and European practice of mounting a pump to its drive motor using a bellhousing. This practice is finding greater usage in America as well.

10.3.1 Plate Mounting

Figure 10.5 shows a typical configuration for isolating a pump on a base. Generally, four identical mounts located symmetrically around the center of gravity are used. The number of mounts can be three or greater, and mount equality is not a requirement for these isolation systems. Where mount centerlines are not equidistant from the

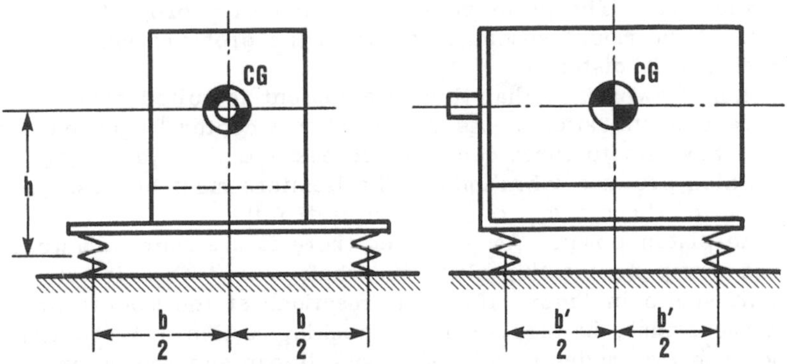

FIGURE 10.5 Typical pump isolator arrangement.

center of gravity, as in Figure 10.6, all static deflections must equal the deflection commensurate with the required natural frequency. Close mounts carry more static load than the more distant ones, so they must have proportionally greater stiffness. The total of all the individual spring rates must equal the spring rate required for the desired natural frequency, as determined with the equations developed in Section 10.2.1.

Although this configuration is often used successfully, it has some serious drawbacks. The easiest to see is that pump location, therefore alignment, changes with shaft torque. When torque is applied to the shaft, it is resisted by the mounts. Those on one side of the shaft are compressed and those on the other extend. The net effect is that the pump rotates about a point midway between the mounts. With stiff mounts this motion may be within the limits allowed by the shaft's flexible coupling. With soft mounts the motion

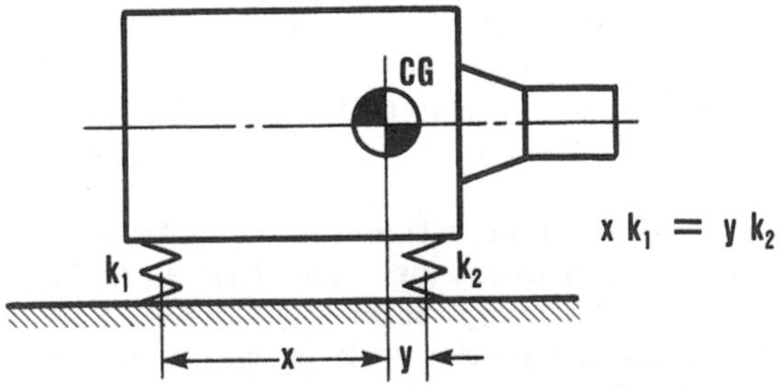

FIGURE 10.6 For equal static deflections, unsymmetrical mounts must have stiffnesses proportional to their loads.

may be excessive. The pump movement is inversely proportional to the square of the mount spacing, so borderline problems are solved by increasing this distance.

The other drawback is that this arrangement is suited best for isolating vertical vibrations. As discussed in Chapter 2, pumps and motors can have up to three conjugate forces and three conjugate moments which produce vibrations. The isolators must be designed to handle all of these that exist in a subject unit.

In inline piston pumps, for example, there is a strong structure-borne noise component acting along the shaft axis. When it acts on the unit shown in Figure 10.5, the reactions at the mounts produce a moment that, in turn, causes a rocking motion. A problem arises because two modes of vibration, one linear and the other rotary, are driven by a single force. The two vibration modes in this case are said to be *coupled.*

Coupled vibration modes have unique and usually undesirable characteristics. Each of these modes, if excited singly, would have its own natural frequency that could be calculated from the equations developed in Section 10.2.1. Further, they would isolate forces and moments as described in Section 10.1. When they are coupled, they are a two degree of freedom system that has two resonant frequencies. Isolation is governed by the highest one. Calculating the natural frequencies of such systems is very complex, but Crede (4) has simplified the process for isolator configurations like that in Figure 10.5.

With dual resonant frequencies, coupled vibration systems need generally wider windows than those present in the structureborne noise spectrum that we were discussing. Although design parameters can be optimized to minimize the spread in the natural frequencies, this is a complex cut-and-try process. It is usually better to go to a mounting configuration like that shown in Figure 10.7, which decouples the modes.

Mode decoupling is accomplished by placing the center of the mounts in a plane through the center of gravity. The same effect is achieved with the arrangement shown in Figure 10.8. With such arrangements, forces along the pump axis produce only translation. Further, while shaft torque still causes pump rotation, the center of rotation coincides with the shaft axis, so no change in alignment occurs.

Shaft torque variations and internal vibrations such as those of the yoke in piston pumps cause structureborne noise that is isolated by tuning the rotational modes of the pump mounting. This is seldom considered in designing pump isolation. It is not known if the success of mounts designed without consideration of these modes is due to the chance optimization of the appropriate parameters or because such modes are usually not important noise sources. Perhaps the following discussion of rotational modes can be ignored for

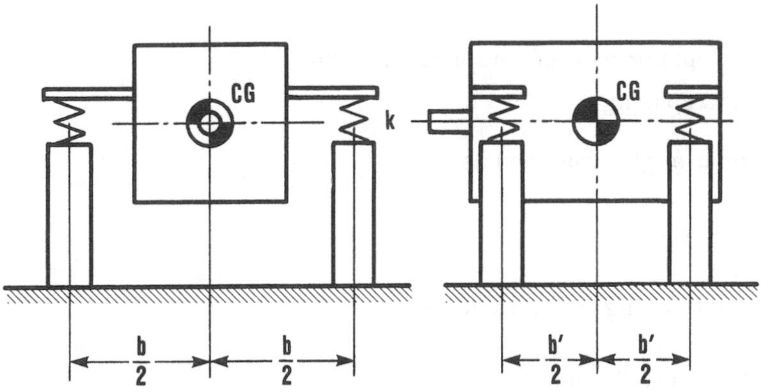

FIGURE 10.7 Vibration modes are decoupled when the center of gravity lies in the plane of the mounts.

most commercial applications. It is included because it may be helpful in critical applications such as those on naval ships.

Sometimes tuning an isolator's rotational modes is difficult. The spring rate for these modes depends on the mount spacing as well as their spring rates. In Figure 10.7 a rotation about the shaft axis that compresses one pair of springs 1 in. and lets the other pair extend the same amount produces a torque of

$$T = \frac{4kb}{2} = 2kb, \quad \text{in.-lb}$$

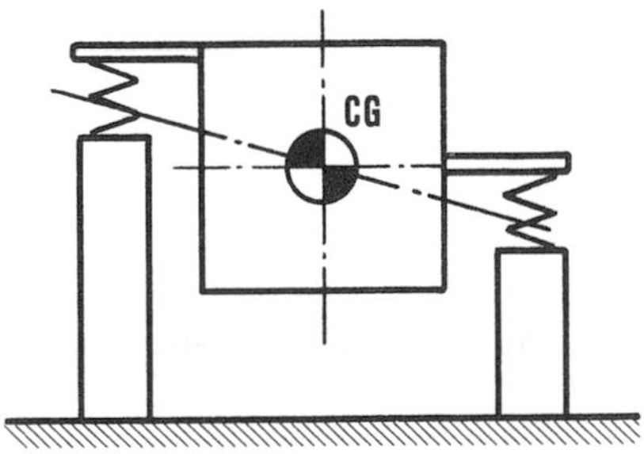

FIGURE 10.8 Alternative configuration for decoupling vibration modes.

where

k = axial spring rate of one mount, lb/in.

b = mount spacing, in.

The rotational angle produced is

$$\theta = \frac{1}{b/2} \quad \text{rad}$$

The rotational spring rate, then, is

$$K_r = kb^2 \quad \text{in.-lb/rad}$$

To be compatible with other equations developed in this chapter, the total mount axial spring is substituted

$$K = 4k \quad \text{lb/in.}$$

so

$$K_r = \frac{Kb^2}{4} \quad \text{in.-lb/rad}$$

Since the mount linear spring rate is selected to provide the required vertical natural frequency, only the mount spacing can be varied to tune the rotational mode. Substituting the foregoing rotational spring rate into the equation for rotational natural frequency, we have

$$f_{nr} = 3.13 \sqrt{\frac{K_r}{wr^2}} \quad \text{Hz}$$

$$f_{nr} = 3.13 \sqrt{\frac{Kb^2/4}{wr^2}} \quad \text{Hz}$$

Comparing this expression with that for the natural frequency for linear modes, f_n, we see that

$$f_{nr} = f_n \frac{b}{2r} \quad \text{Hz}$$

All the linear modes will be tuned to the same frequency if the mounts have transverse spring rates equal to their axial rates. Rotational modes can have the same natural frequency only if their mount spacing is equal to twice the radius of gyration about the axis of the rotation.

There are times when isolators cannot be put where their centers are in a plane with the center of gravity of the supported mass. In these cases, decoupling of the vibration modes can be accomplished by using inclined mounts as shown in Figure 10.9. The mounts lie in planes parallel to the paper and are inclined in only one direction.

This technique only decouples one linear mode from one rotational mode. In the example shown in Figure 10.9, the inclined mounts ensure that a horizontal force passing through the center of gravity produces only linear motion and no rotation. A vertical load, as in the case of parallel mounts, is not coupled because of the symmetry of the mounts about its line of action. The third conjugate load, at right angles to the plane of the paper, is still coupled and produces both linear and rotational motion.

In inline piston pumps the strongest structureborne noise is due to the forcing function acting along the shaft axis. To uncouple modes for this load, the mounts are inclined in planes parallel to the load. There is no horizontal force at right angles to this one, so the fact that a force in this direction causes coupled vibrations is of no consequence.

When parallel mounts are used, their transverse and axial stiffnesses must be equal for equal natural frequencies of all the linear modes. With nonparallel mounts it is necessary to use mounts whose

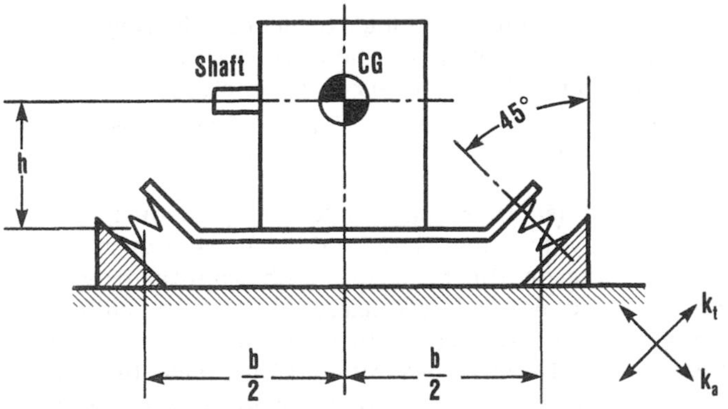

FIGURE 10.9 Angled mounts can also be used to decouple vibration modes.

axial and transverse stiffness are not equal. By inclining the mounts at 45° the effective spring rates in the vertical and horizontal directions of Figure 10.9 are equal. These total spring rates are

$$K = 2(k_a + k_t) \quad \text{lb/in.}$$

where

k_a = single mount axial spring rate, lb/in.

k_t = single mount transverse spring rate, lb/in.

The total spring rate at right angles to these two directions is only due to the transverse stiffness. This rate is

$$K = 4k_t \quad \text{lb/in.}$$

Since transverse stiffness is always less than the axial in these mounting systems, the natural frequency in this direction is lower than in the other two directions.

The design of inclined mounts is complex, but it, too, has been simplified by Crede (5). His procedures are for a very broad spectrum of possible cases, so additional simplification occurs when we consider only pump mountings. Designing inclined isolators is still a cut-and-try process, however.

Decoupling occurs when the elastic axis of the mounts passes through the center of gravity. The location of this axis is a function of the mount spacing and ratio of the axial to the transverse stiffnesses as well as mount inclination. The relationship between the first two of these parameters, in terms of dimensionless ratios, is shown in Figure 10.10. This curve applies to mounts inclined at 45°.

As is the case of parallel mounts, the ratio of the natural frequencies in the rotational and linear modes is a function of the ratio of the mount spacing and the radius of gyration. It is also a function of the mount stiffness ratio. These relationships are shown in Figure 10.11.

There is no set way to use these curves to design a pump mounting. Perhaps the best way to start is to find what mounts might be used. Selection is made on the basis of which mounts provide the desired spring rate and natural frequency for the uncoupled linear mode. Catalogs and vendor contacts provide the spring rates and stiffness ratios. With the stiffness ratios of the selected mounts, the height-to-spacing ratios required by these mounts is determined from Figure 10.10.

A layout of the pump assembly will show which of these ratios are in the practical range and what spacings are likely. The ratio

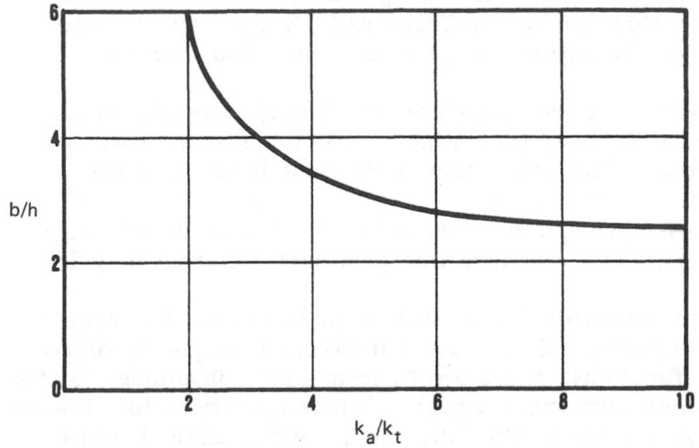

FIGURE 10.10 Spacing requirements for decoupling mounts on a 45° angle. (From Ref. 5.)

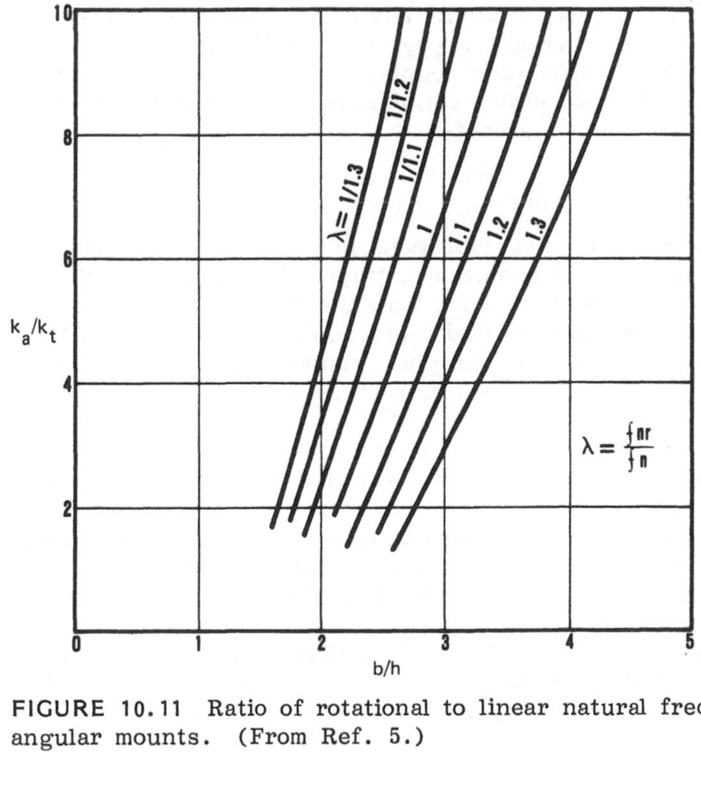

FIGURE 10.11 Ratio of rotational to linear natural frequencies of angular mounts. (From Ref. 5.)

of the mount spacing to the radius of gyration, b/r, can then be calculated. With this and the stiffness ratio, Figure 10.11 determines the ratio of the natural frequencies of the two vibration modes.

Judging what frequency ratios are satisfactory depends on the width of the available frequency window. If the spectrum window was only an octave wide, the rotary mode must have the same natural frequency as the linear. If the window is wider, a difference in the two frequencies is acceptable, the degree depending on the width of the window and how the linear natural frequency is positioned in it.

There is an extenuating factor that often broadens the range of acceptable frequency ratios. The fundamental frequency of most moment forcing functions, therefore their angular vibrations, is the same as the second pumping harmonic. Their spectra is like that of the linear vibrations except that the odd pumping harmonics are missing. The windows for rotary modes, therefore, are often wider than those for the linear modes.

Given a spring rate and stiffness ratio, the only factor that can be varied to adjust the natural frequency of the angular mode is the mount spacing. This, of course, must be done so that the height-to-spacing ratio remains appropriate for the stiffness ratio. Deviating from the specific ratio couples the two modes. In extreme cases where no other acceptable combination is satisfactory, a small degree of coupling may be acceptable.

10.3.2 Bellhousing Mounting

The strategies for decoupling vibrations in plate mountings also theoretically apply to bellhousing mountings. With this type of mounting, however, problems arise from the fact that the pumps are made to be flange mounted at their shaft end. Attempts at putting isolators in the same plane as the pump center of gravity or using angled mounts end up very cumbersome and do not appear to be practical.

Decoupling is not essential to effective isolation. Coupled modes are two-degree-of-freedom systems whose natural frequencies are difficult to calculate. They cause trouble because the highest resonant frequency of such systems is higher than that predicted by simple system calculations. This ceases to be a problem when the natural frequencies are well below 500 Hz, the lower boundary of the critical noise components.

This has been demonstrated by BHRA Engineering with a resilient bellhousing mount that provides very effective isolation without decoupling (6). This isolation system is shown in Figure 10.12. It utilizes four commercially available bushing-type isolators, which are

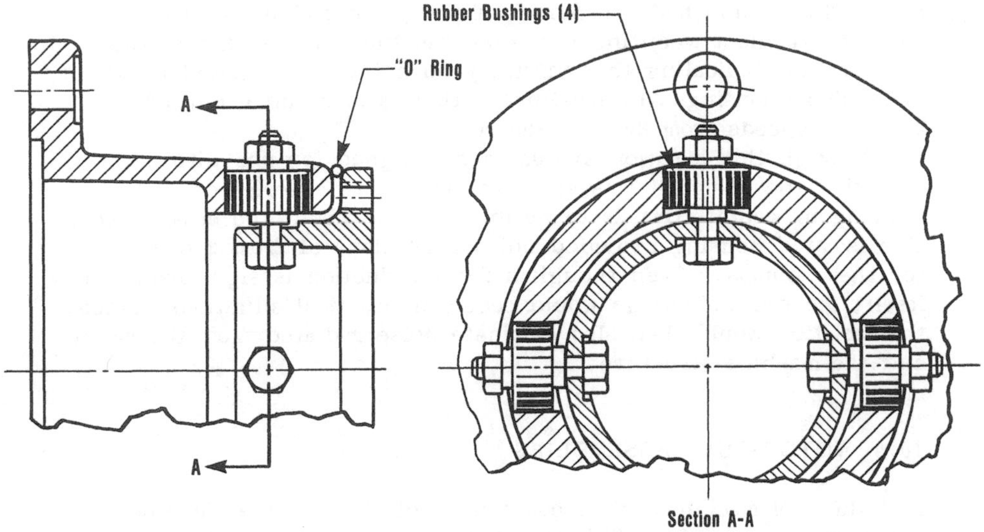

FIGURE 10.12 Isolating bellhousing pump mounting. (From Ref. 6, reprinted by permission of the Council of the Institution of Mechanical Engineers.)

much stiffer radially than they are axially or torsionally. The relatively high overhang moment, is therefore, carried radially by just the top and bottom isolators. The weight itself is carried through the two horizontal isolators. These isolators are located in the plane through the shaft coupling so that their deflections produce a minimum of misalignment at the coupling.

This isolation system was tailored to a specific vane pump. Linear vibrations in the direction of the pump shaft axis are resisted by all four mounts, and these were selected to provide a natural frequency of 38 Hz. The rotational vibrations about this axis are also resisted by all four isolators and had a calculated natural frequency of 52 Hz. The measured natural frequencies were 55 and 64 Hz, respectively. There are two sets of coupled modes, and each of these had resonances detected at 20 and 78 Hz.

All of these frequencies are low in comparison to the critical noise frequencies from 500 to 2000 Hz. The pump was operated throughout the range 800 to 1800 rev/min, so the shaft frequency varied from 13 to 30 Hz. Only the lowest of the coupled resonances falls in this range.

When this system was evaluated, only the noise coming from the pump and bellhousing was measured. Previous tests using the pump for which the isolation was designed found that the pump noise alone was 3 dB less than the total noise from the pump and the bellhousing. This established the parity of the pump and housing noise. With the resilient mounting the total was the same as that of the pump alone, indicating that virtually no noise was radiated by the isolated bellhousing. Interestingly, this same result occurred at all pump speeds from 800 to 1800 rev/min.

Although this isolation system was designed for use with a specific vane pump, it was also evaluated with a series of pumps varying in weight from 24 to 59 lb. All had about the same power. The isolated housing provided noise reductions of from 2 to 6 dB for these pumps. Even though a 2-dB reduction is significant, in itself, greater reductions would occur in practical situations, since the isolation would then also eliminate noise radiated from the motor and the machine structure.

10.4 MOTOR-PUMP ASSEMBLIES

In industrial machines, it is common practice to isolate the whole motor-pump assembly. This has several advantages. The first is that the shaft torque is not transmitted through the mounts, and therefore torque changes do not cause deflections. Another advantage is that more mass is being supported and it is easier to find stable mounts that provide the required static deflections. Mounting configurations like those discussed in Section 10.3.1 (Plate Mountings) are used. Attachment of the mounts is also generally easier with this arrangement than when a pump is isolated by itself.

The disadvantage with mounting the whole assembly is that the drive motor and the structure tying the two units together are not isolated from the pump structureborne noise. In Section 9.2.1 the motor-pump mounting plate was identified as a strong noise source. Other assembly structural elements, such as the risers elevating the pump, have also been found to be good airborne noise radiators. Although electric motors generally are not good at radiating sound, they are larger than the pumps that they drive and therefore are potentially better radiators than these pumps. If the pump alone is isolated, these components are kept from being strong noise sources.

10.5 SHAFT ISOLATION

Torsional vibrations were considered in Section 10.3.1 in regard to their transmission through pump and motor mountings. Here we consider their transmission through connecting shafts.

While drive shaft torsional vibrations seldom cause problems, their isolation deserves some consideration. Isolation is easily achieved by using a coupling with a low torsional stiffness which in combination with the pump rotating group moment of inertia, provides a torsional natural frequency well below the pumping frequencies. Misalignment causes second and fourth shaft harmonic frequencies, so these are best avoided.

The presence of resilient material does not ensure that a coupling is soft. A disk of such material, for example, tends to be torsionally stiff while being flexible enough in bending to accommodate considerable misalignment. Couplings transmitting load through metal springs, on the other hand, may be torsionally quite flexible. As with metal spring mounts, the load path through these couplings must be interrupted by resilient material to make them suitable for noise isolation.

Unfortunately, most coupling manufacturers do not publish the torsional spring rates of their products. These data are often available for the asking, however.

Shafts used to manually control pump stroke are also vibration transmitters (7). These are common in farm machinery, where they control piston-type hydrostatic transmission output speeds. In Chapter 2 it was seen that piston pumps have very strong yoke vibrations. With manual controls, these vibrations travel through several control links and then enter the cab sheet metal. They therefore have ample opportunity to produce loud noise.

Rubber isolators are effective in attenuating this transmission. Figure 10.13 shows the cross section of an automotive engine mount that was used for this purpose. This is one case where a high-production part fits a critical need. In this case it was tried because it was available from an automobile dealer's parts department sooner than it could be secured from a regular mount supplier.

Figure 10.14 shows the control simulation used to evaluate the mount. Figure 10.15 shows the vibration levels in the original setup

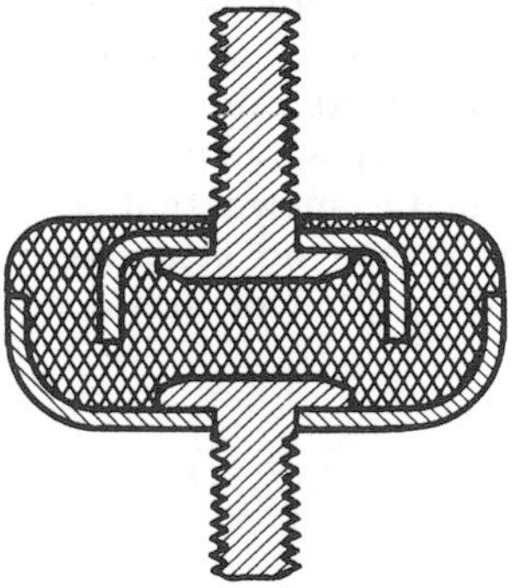

FIGURE 10.13 Engine isolator cross section.

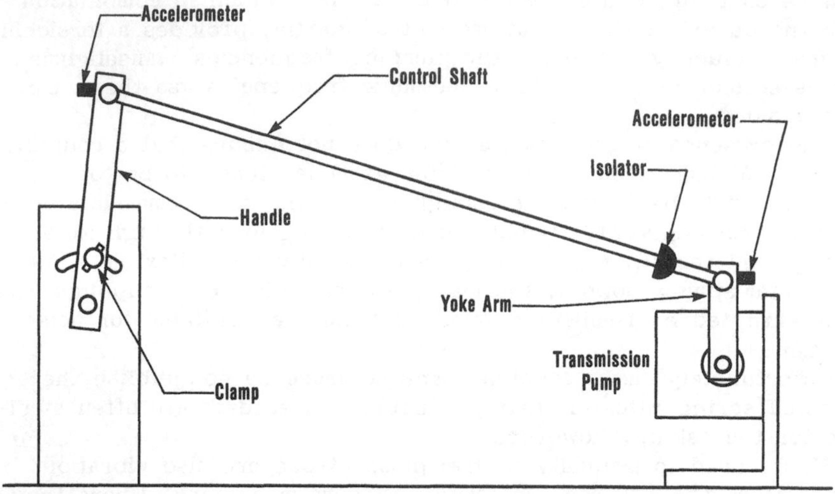

FIGURE 10.14 Pump control simulation for evaluating the control shaft isolator.

and those after the mount was added. The scale of these plots is linear and it can be seen that the mounts reduced the vibrations at the resonant frequency by about 95%. This decreased the noise produced by these vibrations by 26 dB. In practical terms it eliminated that noise source and the noise level dropped to the total level of the remaining sources.

It should be noted that relatively little mass was isolated. There was a concern that if the isolator was soft enough to provide a low natural frequency, the control would be too mushy for the precise control that is required. The natural frequency was not measured but is judged to be fairly high with the stiff mount that was used. However, the strongest yoke vibration component is its fundamental, which is twice the piston frequency. At 1800 rpm this frequency was 540 Hz. In reference to Figure 10.1, it can be shown that a natural frequency of about 120 Hz provides the measured attenuation. This is achieved with less than a thousandth of an inch of static deflection.

10.6 HYDRAULIC-LINE ISOLATION

Hydraulic lines provide one of the most efficient paths for conveying noise energy from its sources to good radiators, such as the machine structure and sheet metal. Both structureborne and fluidborne noise

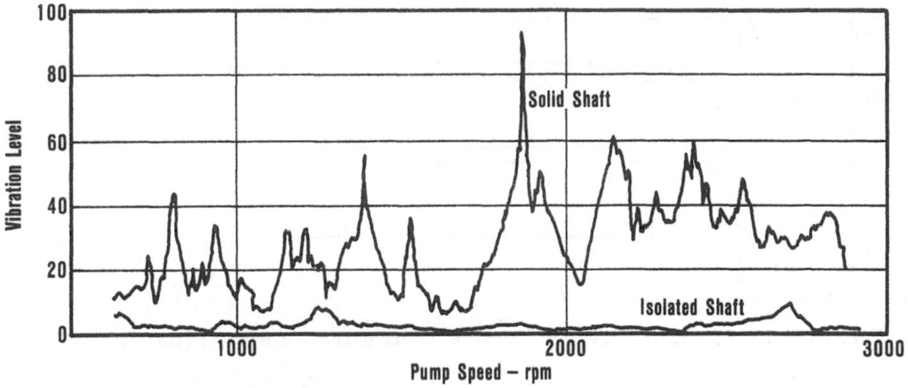

FIGURE 10.15 Effectiveness of control shaft isolator.

are transmitted. Here we are concerned with their transmittal of structureborne noise, in the form of lateral vibration. We are also concerned with the vibrations that result from the action of the fluidborne noise on the lines themselves, as discussed in Section 9.2.2.

There are two types of line isolation; one prevents structureborne noise from reaching the lines. However, since the lines are also a source of vibration, the second type prevents line structureborne noise from reaching good sound radiators. Hydraulic lines are only fair sound radiators, so isolation from more proficient radiators is important. Both types of isolation are required in making a machine quiet.

10.6.1 Flexible Hose and an Isolator

Isolation from noise sources is done with flexible hose. Although structureborne noise attenuation increases with hose length, hose expands radially with pressure, so pressure pulsations cause it to pulse and act as a relatively efficient monopole sound radiator. For this reason, long lines should not consist of hose alone. They should be made up of hose at either end of solid line, as shown in Figure 10.16.

Hose also changes length with pressure, and when bent into a radius it acts like a Bourdon tube, which tries to straighten out with increasing pressure. Both of these actions generate force that causes solid lines to vibrate. The best noise control practice for making bends with hose is to use a solid elbow with two hose sections on either side of it, as shown in Figure 10.17. With this configuration

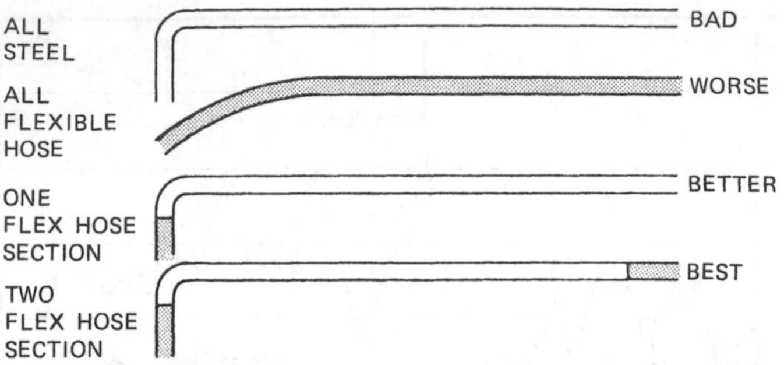

FIGURE 10.16 Long-hydraulic-line configurations. (From Ref. 12.)

there is no Bourdon effect and changes in each hose's length is accommodated by bending of the other hose at right angles to it, so that the transverse force applied to the line is minimal. Figure 10.17 also shows a combination of two parallel hoses that is very effective in providing structureborne noise attentuation without also generating undesirable vibratory forces.

Like other types of vibration isolation, attenuation provided by hose increases with frequency. Valve noise which is all in the high-frequency range is effectively isolated with as little as 0.6 m of hose (8). The researchers that observed this recommended the pumps with their lower frequencies be isolated with 2.5 m of hose. From my experience this seems excessive and probably is based on reducing the pumping frequency and its first few strong harmonics. Since it is usually only necessary to control frequencies above 500 Hz, shorter hoses should be adequate. The line lengths given here

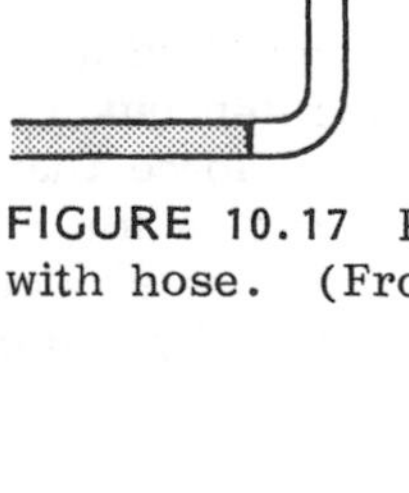

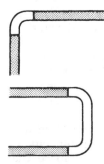

FIGURE 10.17 Preferred methods for making hydraulic line bends with hose. (From Ref. 12.)

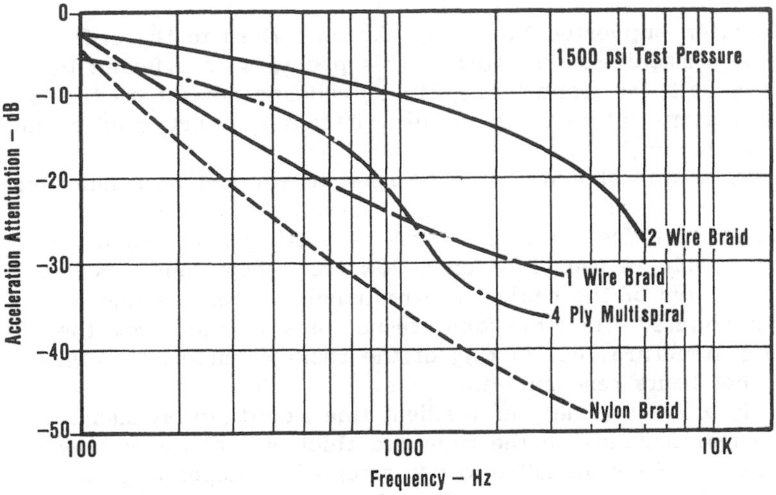

FIGURE 10.18 Structureborne noise attenuation of 1 m of 19-mm hose. (From Ref. 9, reprinted by permission of the Council of the Institution of Mechanical Engineers.)

are for 1-in. four-ply spiral-wound rubber hose. They are overall lengths, with 0.15 m being in the solid end fittings.

Pressure-induced stresses in hose are resisted by reinforcing materials embedded in the hose wall. The reinforcing material and how it is laid affect the ability of the hose in isolating structure-borne noise. Figure 10.18 shows this difference in effectiveness for 1-m lengths of four different hose types. The most effective is the nylon braid plastic hose. Unfortunately, this is only a low-pressure hose. The one-wire braided hose, also a low-pressure hose, was about 70% as effective.

The four-ply multispiral hose was the most effective of the two high-pressure hoses. It was about as good as the single-wire braid. The other high-pressure hose, the two-wire braided, was only 40% as effective as the spiral-wound hose. The poorer performance of the braided hose is believed due to pressure-induced friction between the crossed wires which causes them to lock up at high pressure.

This discussion has been directed toward using hose to isolate lines from noise sources such as pumps, motors, and valves. Hose is also needed on the downstream end of lines to prevent transferring noise to passive hydraulic components that are good sound radiators or that are a good mechanical link to proficient radiators.

10.6.2 Line Support Isolation

Lines are often supported by solidly clamping them to the stiff members of a machine's structure. This provides an efficeint but undesirable path for transferring structureborne noise from the line. Once this energy enters the structure, it excites every good sound radiator in the machine.

Isolation mounts should always be used to support hydraulic lines. This is in conflict with the practice of clamping a line to a rigid member every few feet to keep it from resonating (9). Such restraint only reduces the vibration at low frequencies and ensures efficient transfer of the critical midfrequencies. When a line is resiliently mounted, its vibrations are not only isolated from the supporting structure, but flexing of the resilient material provides damping that limits resonant vibrations.

There is a great variety of resilient line mountings available. Their softness depends on the type and thickness of the resilient material used. They all fall into two categories, depending on the type of resilient material used (10). Those using rubber are usually softer than those using plastic, as seen in Figure 10.19. However, stiffness varies with frequency and that of the rubber peaks in the important midfrequency range. Plastic that peaks at higher fre-

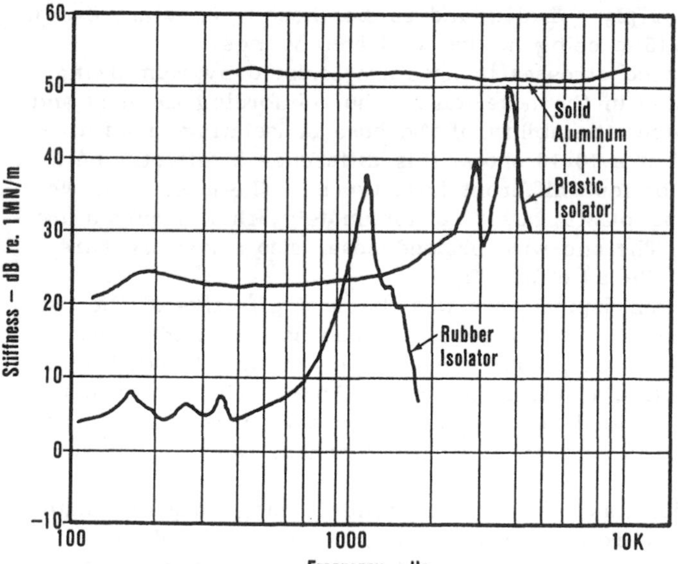

FIGURE 10.19 Typical line support stiffness characteristics. (From Ref. 11, with permission of BHRA, The Fluid Engineering Centre.)

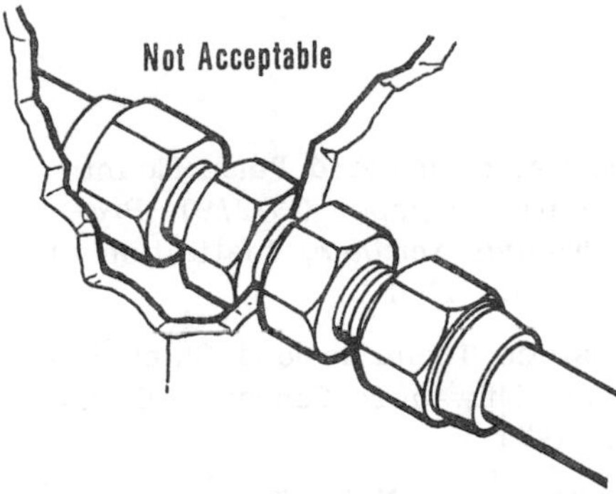

FIGURE 10.20 Bulkhead fittings are the key component for hydraulic loudspeakers. (From Ref. 12.)

quencies, then, tends to be more effective than rubber. Although this is important, it is felt that if either material is thick enough, adequate isolation will be provided.

10.6.3 Isolating Bulkheads

Lines frequently pass through sheet metal panels. The most convenient way to provide for this is to use bulkhead fittings like the one shown in Figure 10.20. This facilitates assembly because the sections of the line on either side of the panel are easily attached to the two ends of such fittings. The trouble with this arrangement is that it constitutes a very efficient hydraulic loudspeaker. All of the structureborne noise of the line is available to the sheet metal, which is an effective sound radiator.

Extraordinary effort must be used to avoid any mechanical contact between lines and sheet metal. Where permitted, the line can just pass through a clearance hole in the panel. In cases where a seal around the line is required, as in the case of enclosures, it requires considerable ingenuity to provide both sealing and easy assembly. This task has been discussed in some detail in Section 9.3.2.

REFERENCES

1. C. E. Crede, *Vibration and Shock Isolation*, Wiley, New York, 1951, pp. 16–20.
2. *Sound and Vibration*, P. O. Box 40416, Bay Village, Ohio.

3. Ref. 1, pp. 203–214.

4. Ref. 1, pp. 43–72.

5. Ref. 1, pp. 73–85.

6. R. A. Heron and J. May, "The Use of Isolated Pump Mounts for Fluid Power System Noise Control," paper C379/80, *Proceedings of the Quieter Oil Hydraulics Seminar,* Institution of Mechanical Engineers, Oct. 1980, pp. 47–53.

7. S. J. Skaistis, "Keeping Hydrostatic Transmissions Quiet," *Hydrostatic Transmission Seminar,* Milwaukee School of Engineering, Milwaukee, Wis., March 1979.

8. R. A. Heron and I. Hansford, "Airborne Noise Due to Structure Borne Vibrations Transmitted Through Pump Mountings and Along Circuits," paper C259/77, *Proceedings of the Quiet Oil Hydraulic Systems Seminar,* Institution of Mechanical Engineers, Nov. 1977, pp. 41–50.

9. M. L. Hughes and B. C. G. Sanders, "The Attenuation Properties of Hydraulic Hose," paper C381/80, *Proceedings of the Quieter Oil Hydraulics Seminar,* Institution of Mechanical Engineers, Oct. 1980, pp. 63–70.

10. M. L. Hughes, "The Dynamic Properties of Resilient Pipe Clamps," paper C386/80, *Proceedings of the Quieter Oil Hydraulics Seminar,* Institution of Mechanical Engineers, Oct. 1980, pp. 87–94.

11. M. L. Hughes and S. R. Ball, "An Investigation into the Vibration Isolation Properties of Some Commercial Pipe Clamps for Fluid Power Systems," Rept. BH 17, BHRA, The Fluid Engineering Centre, Bedford, England, Feb. 1979.

12. *More Sound Advice*, Vickers Inc., Troy, Mich.

11
Reducing Fluidborne Noise

Pump fluidborne and structureborne noises have about the same high-energy levels. They therefore have the same potential for producing loud audible noise.

Since their spectra are very similar, evaluation of the airborne noise provides no clues as to which predominates. From experience, structureborne noise appears to be involved in problem noises more often than fluidborne. On this basis, in quieting a machine, it is recommended that first efforts be directed to vibration control. Generally, reducing fluidborne noise only should be considered after this approach fails to produce the required results.

This does not imply that fluidborne noise is unimportant. In some machines it is the major cause of excessive audible noise, and reducing it is the only way of solving the problem.

11.1 ACOUSTIC FILTERS IN GENERAL

Devices that attenuate fluidborne noise are called acoustic filters to distinguish them from conventional contamination filters. Unlike dirt filters, their performance depends on the details of the circuits in which they are used.

11.1.1 Circuit Interaction

Fluidborne noise reductions resulting from adding an acoustic filter are referred to as *insertion losses*. Downstream acoustic impedances reflect back part of the waves transmitted to them through a filter and this reduces the filter's performance. This influence varies with the circuit parameters and can be determined only for a given filter in a given circuit.

In this chapter we discuss filter *transmission losses*, which are the insertion losses that occur when there are no reflected waves from the downstream circuit. This convention makes it possible to pass on general characteristics of filters. It is understood that the actual attenuations attained in a given circuit are this transmission loss minus the effect of waves reflected upstream by that circuit.

Transmission loss (TL) is defined as

$$\text{TL} = 20 \log \frac{\text{incident wave pressure}}{\text{transmitted wave pressure}} \quad \text{dB}$$

or

$$\text{TL} = 10 \log \frac{\text{incident wave power}}{\text{transmitted wave power}} \quad \text{dB}$$

The following discussion reviews some basic filter units to provide guidance in selecting and designing filters. Commercial acoustic filters generally are made up of combinations or multiples of such units. This is done to broaden the frequency range over which high attenuations are provided. Effectiveness of these combinations are approximated by adding the attenuations, in decibels, of their elements. More exact evaluations are made with the impedance analyses discussed in Section 6.1.3. It is recommended that Refs. 1–3 be consulted in making these analyses or in designing filters. Measured attenuations of some of the commercially available combinations have been published (1) or are available from their makers.

Many filters reduce fluidborne noise in their downstream circuit by reflecting part of their incident waves back upstream. This increases the noise in the upstream circuit. Filters also increase upstream noise when their installation makes the circuit's length one that resonates at a strong noise frequency. If this cricuit is very sensitive to fluid pulsations, adding a filter can actually increase a machine's noise. Because of this possibility, the upstream circuit must be critically examined whenever acoustic filters are considered.

11.1.2 Filter Rating

It is easy to eliminate circuit reflections in making mathematical analyses of transmission losses. It is very difficult to do this experimentally. Determining strengths of incident and transmitted waves is also difficult when measuring filter transmission losses.

Because of the difficulties involved, the National Engineering Laboratory in Scotland has developed filter evaluation into a very sophisticated science (4). They make their tests with single-frequency waves supplied by an electrodynamically driven piston. Measurements are made at 10-Hz increments over the range 100 to 2000 HZ.

Both the generated incident wave and the filter reflected waves are present in the line upstream of the test filter, but only the strength of the incident wave enters the transmission loss equation. A computer analyzes measurements made at up to 20 points along this line, which is 200 ft long, to determine the strengths of the two waves. It also has to evaluate the speed of sound, line sttenuation, and the phase angle between the two waves because these vary with frequency and from test to test with small changes in pressure and temperature.

They found that 200 ft of flexible hose downstream of a filter provides sufficient absorption virtually to eliminate discharge line reflections and simplifies transmitted wave measurements. Since such facilities cannot be justified by most hydraulic laboratories, there must be a high reliance on published and calculated data.

11.1.3 Cost

Acoustic filters tend to be expensive. They are moderately large pressure vessels, and safety considerations lead to heavy construction to provide adequate strength even under fatigue loading. None appear to be in high production. Further, the designs and characteristics of some of the commercial units suggests that they are the product of considerable development effort.

Although their cost seems to be justifiable, it is probably the leading reason why so few are used in machines. Most of their use appears to be in naval ships, where cost is not the primary consideration. The complaint is often heard that they cost as much as all of the other hydraulic components in the system. Because of this, all other noise reduction possibilities are tried before filters are even considered.

Their cost can be reduced. One way is to scale them to provide lower attenuations. Most noise problems require reductions on the order of only 10 dB. Published data on commercial filters indicate that most provide much higher transmission losses; many are in the range 30 to 40 dB, which is 100 to 1000 times greater than needed.

In some cases somewhat higher losses are justified because they broaden the frequency range over which reasonable attenuations occur. Cost is also influenced by this range. Most filters appear to have been designed to provide good losses down to 200 Hz or lower. This may be due to their being intended for naval use, where pumping fundamentals and their strongest harmonics must be controlled. For machine use where only the range 500 to 2000 Hz is likely to be important, they could be smaller and, in some cases, simpler.

It is possible that machine builders may find that they can build their own filters, designed to scaled down specifications, which are economically feasible. In some mobile machines, such as farm tractors, portions of the hydraulic system are an integral part of their structural members. In such machines it may be possible to reduce filter costs further by casting filter cavities into these members rather than making them separate units.

11.2 ABSORPTIVE FILTERS

Fluidborne noise comprises flow perturbation that causes periodically varying amounts of fluid to occupy a given space. Pressure pulsations are the result of the varying amounts of fluid compression, which equal

$$dp = B \frac{dV}{V} \quad \text{psi}$$

where

B = fluid bulk modulus, psi
dV = fluid volume increment, in.^{3}
V = fluid volume, in.^{3}

Absorptive devices reduce pressure pulsations by reducing the effective bulk modulus or by increasing the available volume.

11.2.1 Fluid Volume

The most direct way to reduce the effect of fluid variations is to make the space that receives them larger. Doubling volume, for example, cuts the pressure pulsations in half (6 dB).

Figure 11.1 shows a pump in which volume was added to its discharge port. This is the most effective place to increase volume because it reduces pressure pulsations before they act on the rest of the circuit. Aircraft pumps sometimes have discharge line filters close to them. These filters are intended to trap solid particles but contain a generous amount of fluid. Such volumes are also very effective in reducing fluidborne noise before it acts on the system.

Since fluid waves have finite velocities, the outer regions of large volumes do not react to perturbations at the same time as the region where the disturbance enters. For this reason dimemsions of volume attenuators should not exceed about one-tenth of the wavelength of the highest frequency being filtered. Flat surfaces, which promote standing waves should also be avoided.

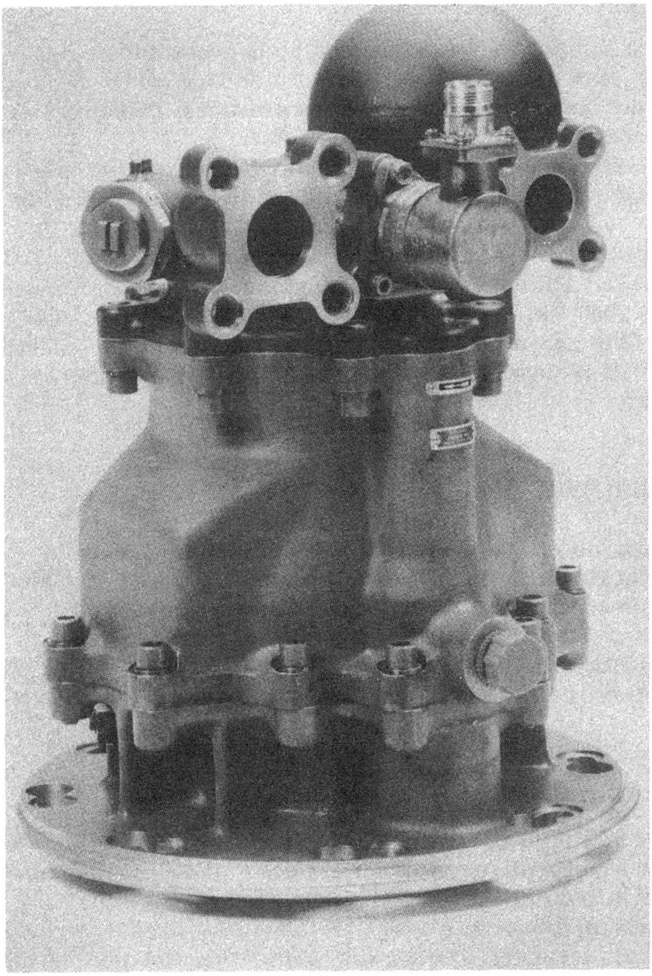

FIGURE 11.1 Increasing a pump's discharge port volume is one of the most effective means of reducing fluidborne noise. (Courtesy of Vickers Inc.)

A sphere is the ideal filter cavity shape. It has the smallest dimensions for a given volume and it has no flat surfaces. Volume attentuators, then, should be made as close to spherical as practical.

These filters have another important requirement. Flow must pass through them. If they are installed as a branch in the circuit, they are a Helmholz resonator and, as we discuss in Section 11.3.4, their attenuation at high frequencies is considerably diminished.

11.2.2 Flexible Hose

Since flexible hose volume increases readily with pressure, it has long been thought that hose is an effective fluidborne noise attenuator. Hose dilation should factor into the previous pressure change equation as a decrease in the bulk modulus, B.

Tests at BHRA Fluid Engineering (5), however, show that this is not true for most hoses. Figure 11.2 shows the results of their tests. Except for low-pressure thermoplastic hose, 2.5-m lengths of 1-in. hose provides practically no attenuation in the critical band from 500 to 2000 Hz. Although high attenuations were measured at frequencies above 2000 Hz, these have little value except in valve noise problems. This accounts for the National Engineering Laboratory having to use 200 ft of hose to produce appreciable absorption in their filter test rig.

11.3 REACTIVE FILTERS

Reactive filters provide attenuations by wave reflection. When a wave reaches a change in acoustic impedance, as discussed in Chapter 6, part of it is reflected and only the remainder is transmitted downstream. The greater the impedance change, the larger the reflection and therefore the smaller the transmitted noise. Automobile mufflers work on this principle and are the most common example of reactive filters.

11.3.1 Expansion Chambers

This filter is shown in Figure 11.3. It is a large cylindrical volume, which differs from a absorptive volume in that its ends are flat and its length is large in comparison to relevant noise wavelengths. Waves are reflected upstream from both its inlet and discharge ports. Maximum attenuation occurs at frequencies where the distance between these reflecting planes equals an odd multiple of their quarter-wavelength. None occurs when the distance is a multiple of a frequency's half-wavelength.

Figure 11.3 shows these attenuation characteristics. The frequency scale in this figure was normalized by dividing by the design frequency. This is the frequency whose wavelength is twice the chamber length. Attenuations, which are determined by the ratio of the chamber cross-sectional area to that of the line, are high over a fairly broad frequency range. The maximum transmission loss of an expansion chamber filter is

$$\text{maximum TL} = 20 \log \frac{\text{chamber cross-sectional area}}{\text{line cross-sectional area}} - 6 \quad \text{dB}$$

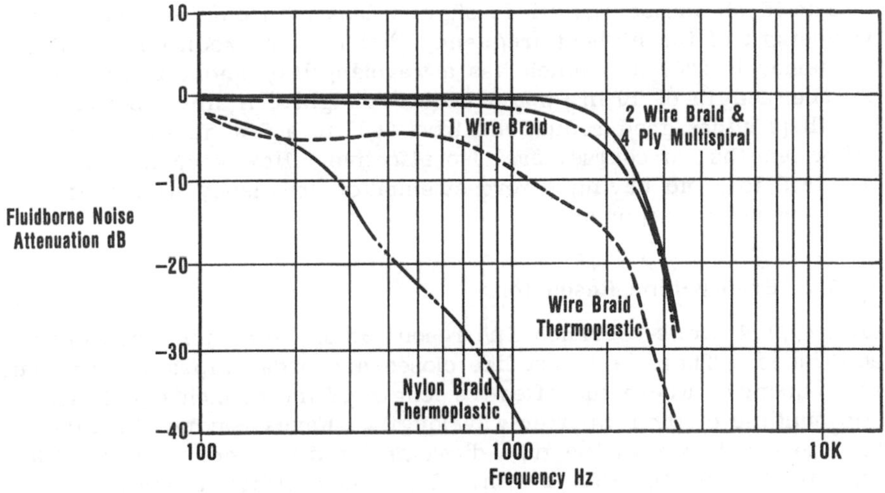

FIGURE 11.2 Fluidborne noise attenuation by 2.5 m of 1-in. hose. (From Ref. 5, reprinted by permission of the Council of the Institution of Mechanical Engineers.)

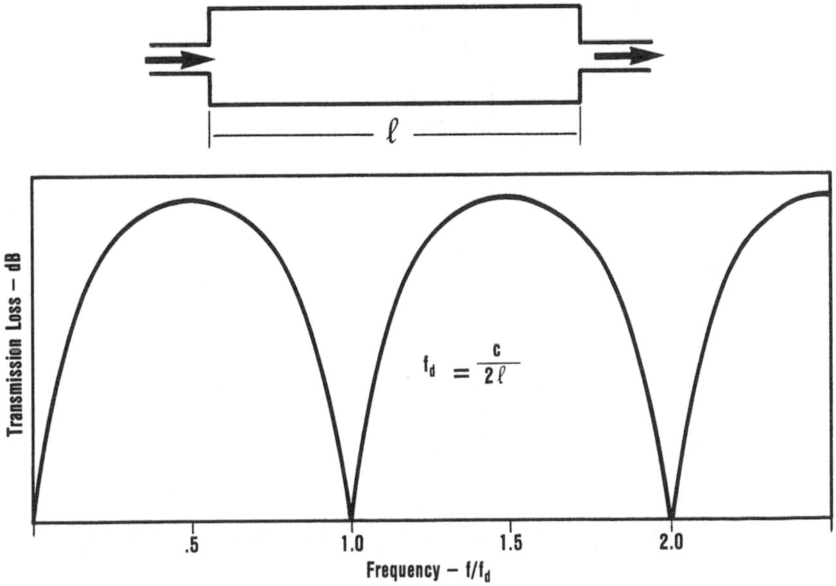

FIGURE 11.3 Expansion chamber filter and characteristics.

The ideal chamber length is slightly less than one-half the wavelength of the highest frequency that must be reduced. If this frequency is 2000 Hz, which has a wavelength of about 23 in., a chamber length of 10 in. would be about right. A filter of this length provides good reductions from 2000 to a few hundred hertz. Other lengths, of course, are also effective. However, care is required to avoid having a zero attenuation frequency near that of a strong noise component.

11.3.2 Branch-Line Resonators

Figure 11.4 shows a branch-line resonator and its attenuation characteristics. These units act like closed-end organ pipes and resonate at frequencies where the effective length of the branch equals an odd multiple of their quarter-wavelength. Waves leaving the tube tend to remain within the tube diameter for a distance of about 0.8 diameter beyond the tube mouth. The tube's effective length, then, is greater than the measured length by this amount. The resonant frequencies of the branch line are

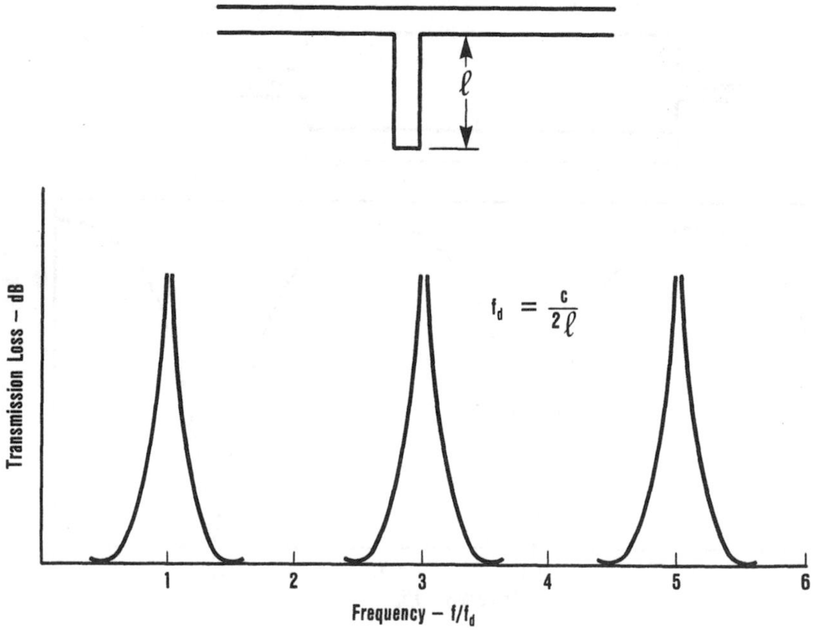

FIGURE 11.4 Branch-line resonator and characteristics.

$$f_n = n \frac{c}{4\ell} \quad \text{Hz}$$

where

n = any odd integer
c = speed of sound in fluid, in./sec
ℓ = effective tube length, in.

At resonance the reflected wave is 180° out if phase with the incident wave, so complete cancellation occurs. Since these devices only cancel the odd harmonics of their design frequency, a single branch does not eliminate all of the pump's fluidborne noise.

It is possible to eliminate, say, the first 10 components of a pump noise by adding more branches. The first branch tuned to the pumping fundamental eliminates the first, third, seventh and ninth harmonics. Tuning the second branch to the second pumping harmonic eliminates the second, sixth, and tenth. It then takes additional branches tuned to the fourth and eighth to complete the job.

Even with four branches, this is an inexpensive filter for eliminating the first 10 harmonics of a pump noise. However, they only attenuate when their resonant frequencies exactly coincide with noise frequencies. Operating changes in pressure and temperature frequently shift the resonant frequencies. Further, pump drive motors change speed with load, so noise frequencies also shift during operation. These variations are sufficient to disrupt the filter action under some conditions. The use of branch filters, is therefore very limited.

11.3.3 Quincke Tubes

Quincke tubes theoretically have the potential for eliminating all harmonics of a pump fluidborne noise, so they are often proposed as a solution for noise problems. However, some of their attenuations have narrow bands like those of branch resonators. For this reason they, like branch resonators, are rarely used. However, they continue to be mentioned so frequently that one wonders if the name has become the generic name for all reactive filters. The purpose of including them here is to put them into perspective.

Quincke tubes filter in two ways. As seen in Figure 11.5, these devices split the flow into two unequal-length paths. Waves traveling through them arrive at the downstream junction out of phase with each other. When the length difference is an odd multiple of a half-wavelength, the phase difference is 180° and complete cancellation occurs. Some cancellation occurs with all phase differences, so the attenuations occur over a relatively wide band. Like the branch resonator filters, however, only the odd multiples of this fundamental resonant frequency are fully attenuated.

The device also filters all multiples of the frequency whose wavelength equals the total length of the two paths (6). The mechanism that does this is not clear, but since it produces very narrow attenuation bands it is concluded that it involves organ pipe resonances. Interestingly, if any of the attenuation frequencies of the two filter modes coincide, they cancel each other and no attenuation occurs at that frequency.

11.3.4 Helmholtz and Chamber Resonators

Figure 11.6 shows both the Helmholtz and chamber resonators. Although they are configured differently, they are acoustically identical. The attenuation curve in the same figure applies to both.

Like branch resonators, these devices utilize the phase shift that occurs in resonance to affect attenuations. They do not have the reoccurring resonances of the branch lines, but they have much wider attenuation bands.

Their operation is similar to that of mechanical isolators. Below their resonant frequency, they behave like the absorptive volumes discussed in Section 11.2.1. Attenuation in this range is determined by the volume. Attenuations above resonance are due to the mass of the fluid in the neck of the Helmholtz resonator and in the holes in the chamber resonator. Since they provide little attenuation at low frequencies, they are sometimes characterized as *low-pass filters*. This type of filter behaves like the pump discharge port that was discussed in Section 6.2.4. As given in that discussion, the resonant frequency is

$$f_n = \frac{\omega_n}{2\pi} = \frac{c}{2\pi}\sqrt{\frac{A}{V\ell}} \qquad \text{Hz}$$

where

ω_n = circular resonant frequency, Hz
c = speed of sound in fluid, in./sec
V = resonator volume, in.^{3}
A = neck cross-sectional area or total area of holes, in.^{2}
ℓ = effective length of neck or holes, in.
= actual length + 1.5d
d = neck or holes diameter, in.

Neglecting friction, these filters have a transmission loss of (7)

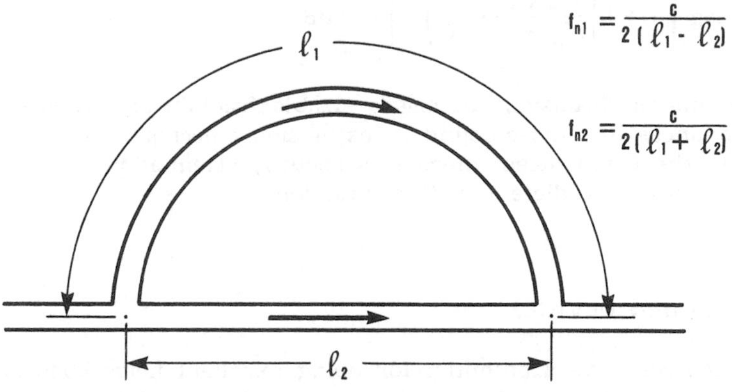

FIGURE 11.5 Quincke tube filter.

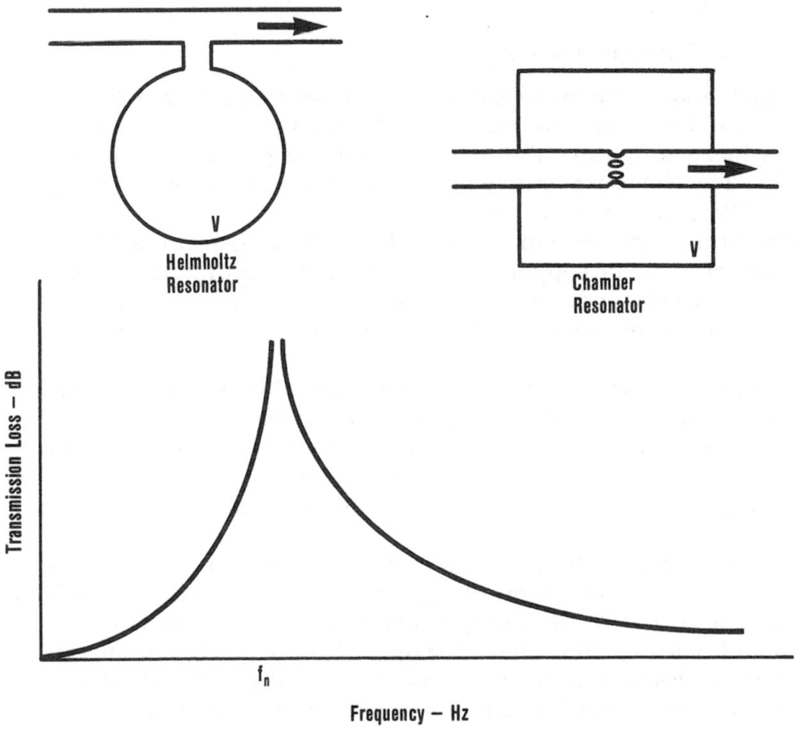

FIGURE 11.6 Helmholtz and chamber resonators.

$$TL = 10 \log \left[1 + \left(\frac{(1/2)(V/\ell A)}{f/f_n - f_n/f} \right)^2 \right] \quad dB$$

Where resonator dimensions become a sizable fraction of a wavelength, resonances like those found in expansion chambers are initiated. At the frequencies where this occurs, attenuations are different than those predicted by this equation.

11.4 GAS-LOADED DEVICES

Gas-loaded devices have a potential for being excellent noise absorbers. Gas has a bulk modulus equal to its pressure. Even at 5000 psi it absorbs pulsations 40 times better than oil. However, structural provisions for maintaining a gas volume within the fluid cause resonances that make some of these devices act more like reactive filters.

11.4.1 Pass-Through Filters

Figure 11.7 shows the construction of a pass-through gas-loaded filter. A rubber tube separates the oil and gas. As the fluid pressure increases, the tube expands and the gas volume is compressed. The volume of gas existing at operating pressures depends on the unit's initial charge pressure.

Whenever system pressure is less than the charge pressure, the tube contracts and is supported by the inner perforated sleeve. No filtering occurs under this condition. The charge pressure, then, must be lower than the lowest expected system operating pressure.

Most of the attenuation provided by these devices comes from absorption. Because of this, they provide good reductions over nearly all frequencies. They have the features of chamber resonators, so they also have resonances. This accounts for the variations in the transmission loss curve of one of these filters, shown in Figure 11.7.

The weak point in these filters is the membrane separating the gas and fluid. At shutoff or whenever the system pressure is less than the charge pressure, the gas tries to force this membrane through the holes of the inner sleeve. If the charge pressure is reduced to decrease stress on the membrane, the noise absorption is decreased along with the gas volume at operating pressure.

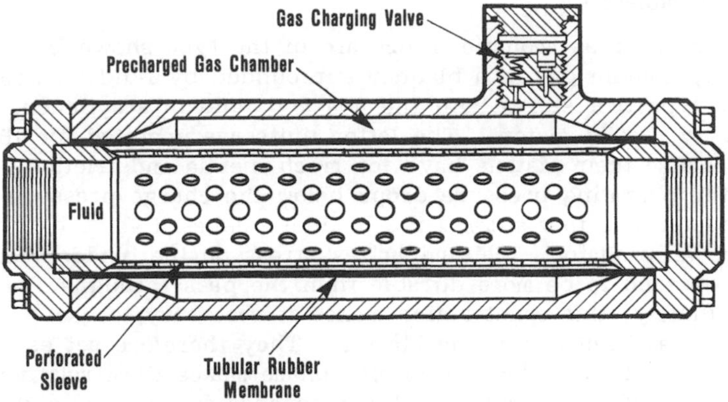

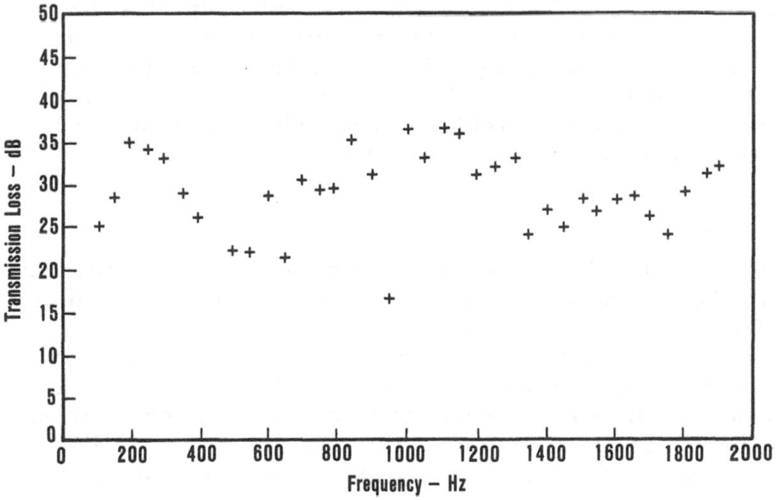

FIGURE 11.7 Pass-through gas-charged filter and measured transmission losses. (From Ref. 1, reprinted by permission of the Council of the Institution of Mechanical Engineers.)

11.4.2 Accumulators

Accumulators used as acoustic filters are of the type shown in Figure 11.8. They contain gas in a bladder surrounded by fluid. There is another type that stores both the gas and fluid in a cylinder, separated by a solid piston. The latter units are unsuited for filter service because their pistons have too much inertia and friction.

When the operating pressure drops below the charge pressure in an accumulator like that shown, the bladder pushes a valve closed to seal the accumulator. This minimizes stress on the bladder, so these filters tend to be more durable than the pass-throughs.

Unfortunately, this construction makes it necessary to use accumulators as branches of the circuit. They therefore act as Helmholtz resonators. Their large air volume makes them relatively soft. Because of this, they have low resonance frequencies and provide little loss in the range 500 to 2000 Hz, as shown in Figure 11.8.

Since they are good at low frequencies, they usually are advertized with oscilloscope traces showing dramatic reductions in pressure pulsations. This simply confirms that the drop is only in the low pumping harmonics. Eliminating noise in the critical range does not make a discernible change in oscilloscope traces. Significant machine noise reductions would be possible if these units were made much smaller, with higher resonant frequencies.

11.4.3 Gas Diffusion

Rubber membranes and bladders do not completely separate the fluid and gas in these devices. A surprising amount of the gas diffuses into the fluid when both are under pressure.

This was discovered by the author when using a bank of very large nitrogen-charged accumulators as a pulseless supply for flow noise measurements. By accident, the system was left pressurized over one night. In the first run on the following day, the oil turned into foam as it passed through the test orifice. Apparently, since the accumulators contained both oil and nitrogen at 2000 psi, the oil became saturated with the gas at that pressure.

Diffusion occurs slowly enough to avoid gas saturation problems during normal operations. Usually, it is not practical to release the charge when a system is shut down, so some outgassing problems can be expected at startup.

Diffusion over long periods depletes accumulator gas charges. They must therefore be monitored constantly. Further, membranes and bladders deteriorate, so periodic checks and replacements are required. The greatest disadvantage of gas-charged filters, then, is that they require appreciable maintenance whereas other filters do not.

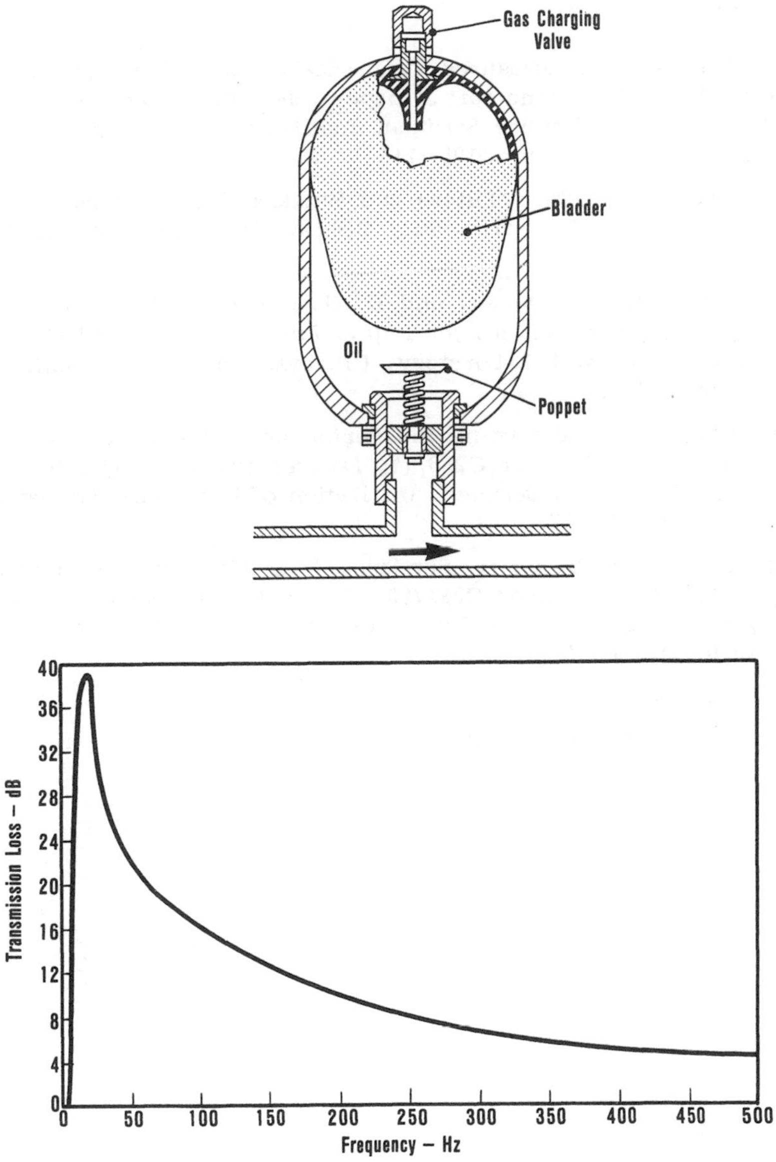

FIGURE 11.8 Gas-charged accumulator and its characteristics as an acoustic filter.

REFERENCES

1. R. J. Whitson, "The Measured Transmission Loss Characteristics of Some Hydraulic Attenuators," paper C382/80, *Proceedings of the Quieter Oil Hydraulics Seminar*, Institution of Mechanical Engineers, Oct. 1980, pp. 105–115.

2. D. D. Davis, "Acoustical Filters and Mufflers," in *Handbook of Noise Control*, C. M. Harris, ed., McGraw-Hill, New York, 1957, Chap. 21.

3. E. L. Hixon, J. V. Kahlbau, and D. G. Galloway, *Guide to the Selection of Acoustic Filters for Liquid-Filled Systems*, DRL-A-202, Defense Research Laboratory, University of Texas, Austin, Texas, Sept. 1962.

4. A. R. Henderson, "Measuring the Performance of Fluid-Borne Noise Attenuators," paper C256/77, *Proceedings of the Quieter Oil Hydraulic Systems Seminar*, Institution of Mechanical Engineers, Nov. 1977, pp. 15–27.

5. M. L. Hughes and B. C. G. Sanders, "The Attenuation Properties of Hydraulic Hose," paper C381/80, *Proceedings of the Quieter Oil Hydraulics Seminar*, Institution of Mechanical Engineers, Oct. 1980, pp. 63–70.

6. Ref. 3, p. 121.

7. Ref. 2, Chap. 21, pp. 22–23.

12 Designing and Developing Quiet Machines

The cheapest quiet machines are those that initially were planned, designed, and developed to be quiet. Many noise controls can be incorporated into a new design for little cost but are expensive when added to an existing machine. This is usually because space restrictions limit the options that can be used. This chapter is directed to making designing and developing a quiet machine more effective and less costly.

Much of the discussion in this chapter applies equally well to the task of quieting an existing machine.

12.1 PLANNING

The importance of planning a new machine is well established in industry. It is equally important in initiating a program for quieting an existing machine.

This planning generally details performance and cost goals and may even consider such topics as size and appearance. Quieting has become such an important performance parameter that it should also be included in such plans.

12.1.1 Setting Noise-Level Goals

The importance of setting noise-level goals for quieting projects was discussed briefly in Section 1.9. It is equally important to set these goals for machines while they are still in their planning stage.

Having a clear noise-level goal saves time in making other planning decisions. For example, it narrows the array of design options being considered. If it indicates the need for noise controls such as isolators or enclosures, other planning factors can be adjusted to accommodate these without losing time.

Goal setting must be done with great care. In the case of an existing machine, the goal is usually based on its measured noise levels under various operating conditions. For new machines, the measured noise levels of similar competitors' products may provide this base. Because of its importance, a thorough survey of how well existing product noise levels meet the needs and expectations of the market should be used.

Customer's purchase specifications often are a starting point in setting noise goals. Most of these are traceable to OSHA imposed work-space noise level limits. They also consider that a number of machines operate in a work space, so the level produced by a single machine must be less than the 90 dB(A) legal limit.

To estimate how much to allow for multiple machines, a group of four identical machines, arranged in a square, are usually assumed. Sound levels from the four machines add together in the work space in the center of the square. From Section 4.3 we know that the level there is 6 dB(A) higher than the level of one machine. Noise levels decrease with the square of distance, so machines beyond the four do not add significantly to this level.

Machine purchase specifications, therefore, commonly place a limit of 84 dB(A) on a machine's noise level. In some cases 85 dB(A) is allowed. Some companies, in anticipation of a drop in OSHA limits, restrict machine noise levels to 80 dB(A). To generalize, noise levels of industrial and mobile machines should not exceed 85 dB(A), and if their level is below 80 dB(A), more companies will consider buying them.

Competitive pressure may impose even more stringent noise limits. This is because quietness is sometimes equated with quality. This was seen when a small Japanese-built plastic molding machine was introduced in this country. Comparable American machines already on the market had noise levels well below OSHA requirements. However, the new machine had been engineered for noise control and was quieter. This gave it a marketing advantage that caused at least one other company to redesign its machine for lower noise.

12.1.2 Estimating Noise Levels

A machine can have almost any noise level, from a very high one to a very low one, depending on how much noise control is built into it. In planning a new machine, we estimate its noise level for some known degree of noise control. With existing machines, of course, this information comes from actual noise measurements. Comparing the existing or estimated noise levels with the goal indicates the extent of the needed noise control effort.

The leading noise generator provides a basis for estimating a proposed machine's noise level. Usually, it is a pump, but it could just as well be a hydraulic motor or, in rare cases, a valve. Generally, its sound pressure level, under the anticipated machine operating conditions, are available from the manufacturer. If not, it must be measured. In the few cases where more than one strong noise source operate simultaneously, their levels should be added.

The exact operating conditions during measurements should be determined for any data that are used. Some manufacturers quote levels at low pressure or even zero flow in their advertising. Use only data measured at within 500 psi, 100 rev/min, and 10% flow of the anticipated operating conditions. Inlet test pressure should be at least a few inches of mercury below atmospheric. Where there is a choice, use data for more rather than less extreme conditions.

A-weighted sound *power* levels of pumps and motors are measured in several different ways, as described in Section 4.4.2. It is customary to give these devices noise ratings in terms of sound *pressure* that is calculated from the measured power. This rating, as described in Section 4.4.3, assumes that sound radiates uniformly in a hemispherical pattern, and the pressure at 1 m from the source center is calculated (1,2). Actually, this is done by subtracting 8 dB from the sound power level. Data that are not labeled explicitly as to whether they represent power or pressure at some distance cannot be used.

The accuracy of directly measured pressures at 1 m depends on the uniformity of the sound field. They are generally considered to be somewhat in error, so if this type of data is all that is available, it is prudent to add several decibels for conservatism.

The noise source's sound pressure rating is useful because it approximates numerically the noise level that the unit will generate in a machine that utilizes normal machine design practices and does not have any special noise control features. Machines having slightly higher levels generally can be brought in line with little effort.

Machine structures radiate noise, so the total audible noise from the machine is greater than that from the noise source alone. However, machine noise pressures are measured 3 ft from a machine's boundary (3). This is farther from the pump or motor than the 1 meter used in their rating. Adding the structure-radiated noise to the lower pump or motor level at this increased distance about equals the rating. This only explains how the two different pressures can be equal. Actually when they are, it is only a coincidence and cannot be justified by scientific arguments. Nevertheless, this chance relationship serves as a useful rule of thumb.

Sometimes crude estimates of machine noise can be refined with judgments based on measurements made on similar machines. Even without such improvements the estimates are generally adequate for planning purposes.

If the estimated or measured level is within a few decibels of the target, it is expected that only simple noise control measures such as reducing the efficiency of some structural radiators is all that will be required. Finding a quieter pump or motor might also provide an adequate noise reduction.

When the difference is 6 to 12 dB, mechanical isolation of the noise source and hydraulic lines are probably required. Enclosures must be considered when greater reductions are needed.

12.1.3 Scheduling

There are many machine details that can affect noise, but only a few need to be designed properly to make the machine quiet. The critical factors vary from machine to machine, so it is not practical initially to engineer everything that could be critical. A machine with all possible noise controls would probably be too expensive. Generally, the course that is followed is to design a machine incorporating some of the noise controls implied by the plan and then let tests of the prototype machine indicate what further noise control efforts are needed. This system works well when enough time is scheduled for the testing and remedial work.

Many machine quieting programs fail because not enough time is allocated for their execution. When time is short, there is great pressure to try quickie tests and fixes. These almost always end up wasting critical time. There is also a tendency to ignore ideas that have long execution times but a high probability of success.

The time that should be scheduled for a machine quieting program depends on the degree of quieting that must be attained, available personnel and facilities, and rework capabilities. It not only varies from company to company but from program to program within a company. Fortunately, estimates improve with experience. About the only advice that can be given to organizations without experience is to allow from two to three times their best estimates.

12.2 PUMP AND MOTOR SELECTION

Since pumps and motors are the leading noise sources, it follows that selecting the quietest ones is the first step in producing a quiet machine. Although selecting a quiet pump or motor is important, it is a mistake to believe that it is the only step required or even that it is the most important step in developing a quiet machine.

Noise differences between pumps of similar capabilities seldom exceed 6 dB(A). The way that a pump is installed in a machine can affect noise levels by 20 dB(A).

Unit selection has to be on the basis of airborne noise ratings. There is not enough data on structure and fluidborne noise for making comparisons. It is generally assumed that steps that reduced audible noise of a pump also reduced the other two noises as well. This is a reasonable assumption when comparing pumps and motors of the same type but may not be valid when comparing different types.

12.2.1 Operating Parameters

Size, speed, and pressure all affect noise, so selecting the optimum combination of these three parameters is the first step in finding a quiet pump. Figure 12.1 shows how pump noise is affected by these parameters; motors are assumed to have the same relationships. It can be seen that speed has the greatest influence. This is because,

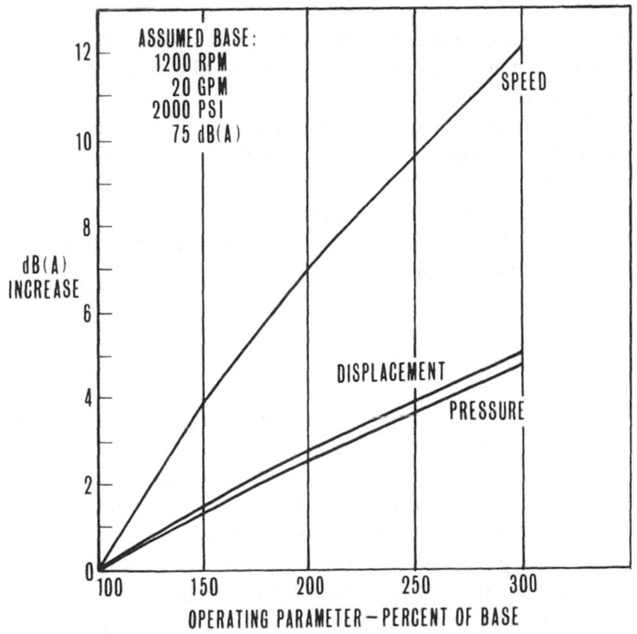

FIGURE 12.1 Effect of changing pump size, speed, and pressure on audible noise. (From Ref. 4.)

as speed increases, more of the strong, lower pumping harmonics move up into the frequency range where they are radiated efficiently.

An engineer in the machine tool industry once made the observation that reducing pump speed was one of the cheapest ways to reduce noise. This came with the discovery that 1200- and 1800-rev/min motors of the same horsepower cost about the same. The cost of decreasing speed, then, was due only to using a larger-displacement pump. This idea was quickly adopted, and today almost all machine tool and injection molding machine pumps in the United States operate at 1200 rev/min.

Noise increases equally with both pressure and pump size, so there is no advantage of trading off one against the other. For quietness, then, the optimum operating parameters are the slowest practical speed and any combination of pressure and displacement that provides the needed hydraulic power.

12.2.2 Pump Types

It is impossible to recommend one pump type as being quieter than others. As an example, internal gear pumps appear to be quieter than almost every other pump type. However, this is true only in the small sizes. In larger sizes other pumps have lower noise levels.

Screw pumps also have a reputation for being the quietest. However, they are hard to compare to other pumps since they have unique speed and fluid requirements.

Only two generalizations about pump noise appear to be valid. One is that the development effort invested in quieting a pump is more important than its type. The other is that fixed-displacement pumps are quieter than variable-displacement ones. Selecting a quiet pump, then, has to be done by comparing the complete specifications and noise levels of all candidate pumps.

12.3 CAVITATION CONTROL

The quietest pump in the world becomes extremely noisy if fed fluid with a lot of entrained air. The mechanics of this increased noise is discussed in Section 5.2.4. Extreme cases make a pump sound like it is crushing marbles. The cause of such cases is quickly recognized and corrected. However, even very slight cavitation raises noise levels appreciably. The increase is not as obvious, so the cause is not readily recognized and the higher noise level is usually attributed to the pump itself. Our concern is that the cause remains uncorrected.

12.3.1 Reservoir Design

Reservoir design is critical in avoiding entrained air. It starts with directing the fluid's return to the reservior. The fluid must never be allowed to pour into the reservoir as shown in Figure 12.2. Such action causes air entrainment. To avoid this condition, the end of the return line must always be below the fluid surface. The drop in this surface caused by the maximum extension of cylinders, the tilting of mobile machines, and infrequent leakage makeup must be considered in locating the end of the return line.

Fluid sometimes returns to the reservoir as high-velocity jets that cause whirlpools and other surface disturbances which entrain air. Diffusers, like that shown in Figure 12.3, break these jets into many low-velocity jets that do not cause these problems. To be effective, the total diffuser hole area should be large enough to reduce the fluid velocity to less than 2 ft/sec.

Another source of entrained air occurs only in off-the-road vehicles. As these vehicles tilt, their fluid sometimes sloshes over the reservoir baffles and captures air bubbles, as shown in Figure 12.4. This condition is eliminated by providing the equalizing holes, which are also shown.

These precautions do not eliminate entrained air; they only minimize it. A great deal of air is already in the returning fluid as a result of throttling in the hydraulic system. One of the main purposes of the reservoir is to eliminate this air. The process used is to keep the fluid in the reservoir long enough for the bubbles to rise to the surface and disperse. Fluid velocity must be kept low during this period to promote separation.

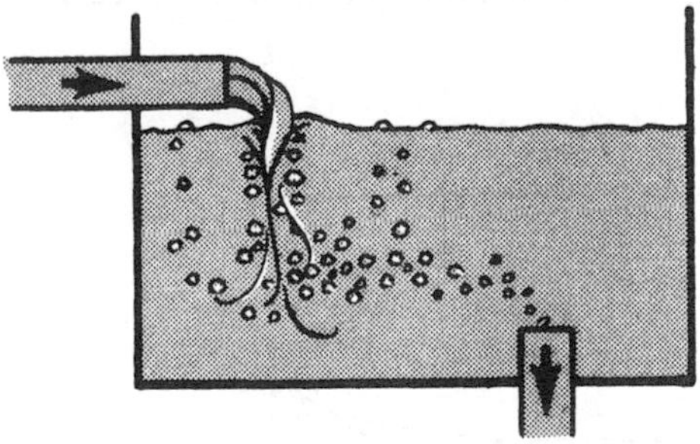

FIGURE 12.2 Fluid returning above the reservoir surface entrains air. (From Ref. 4.)

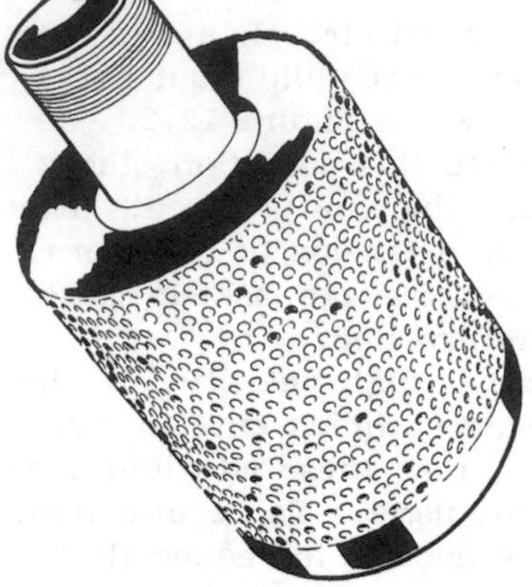

FIGURE 12.3 Return-line diffuser. (From Ref. 4.)

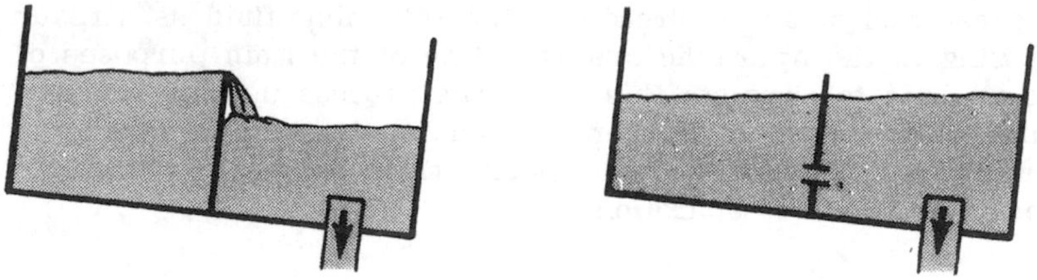

FIGURE 12.4 Equalizing holes in baffles reduce sloshing. (From Ref. 4.)

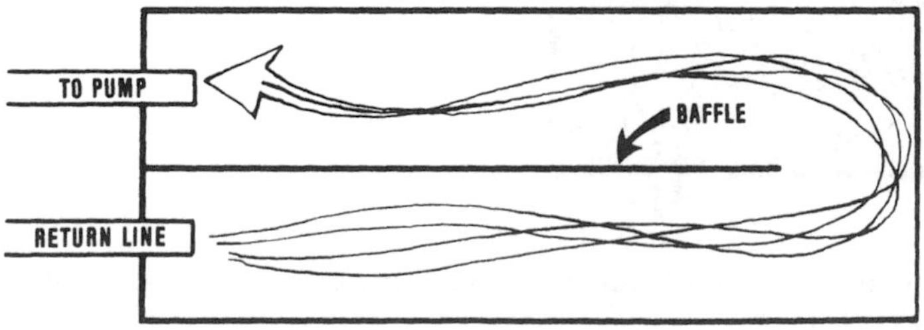

FIGURE 12.5 Reservoirs have baffles to aid bubble release. (From Ref. 4.)

A rule of thumb for sizing industrial machine reservoirs is to provide enough volume for 2 or 3 minutes of the system's maximum flow rate. Even with reservoirs of this size it is necessary to provide baffles, as shown in Figure 12.5, to increase the fluid path and maximize dwell time.

Screens can also be used to aid deaeration. Figure 12.6 shows a scheme that has been used successfully. This figure also shows the effect of mesh size and screen angles. These relationships hold only for the particular reservoir used in the tests; other experimenters report slightly different results. If screens are considered for a machine, it is suggested that various combinations be tried to see what works best for that reservoir.

Suction-line inlets also promote air entrainment. When they are near the surface they do this by forming whirlpools, as shown in

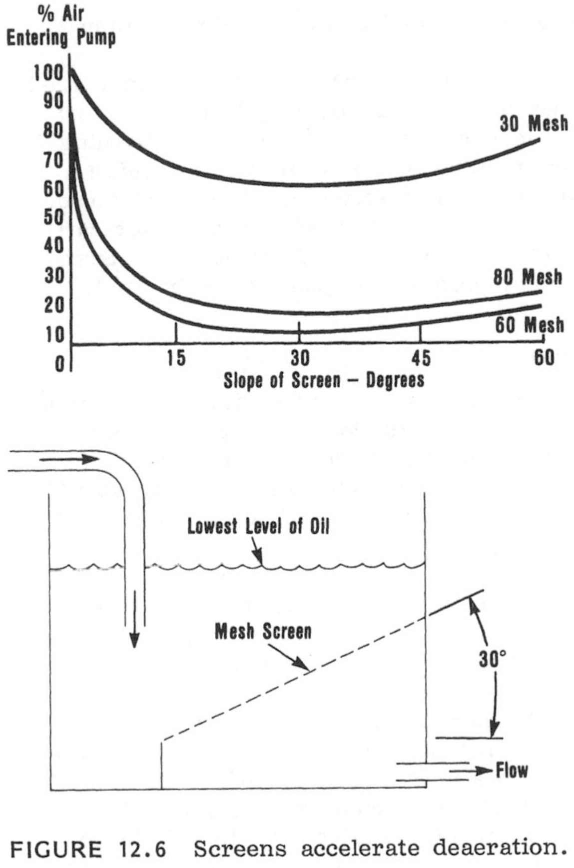

FIGURE 12.6 Screens accelerate deaeration. (From Ref. 4.)

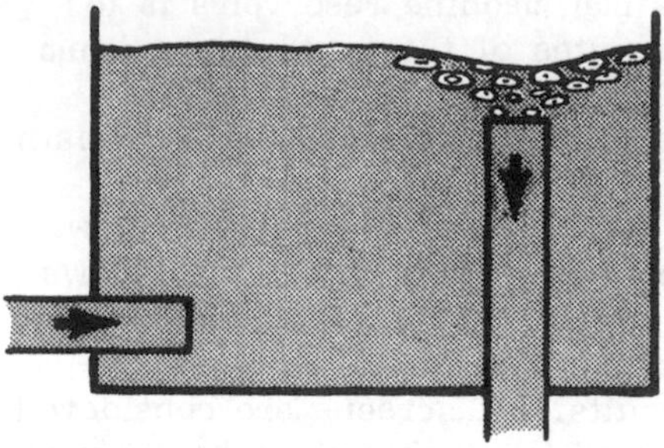

FIGURE 12.7 Suction-line entries near fluid surface ingest air. (From Ref. 4.)

12.7. They must therefore be located at greater depths than those required for return lines.

Mobile machine reservoirs are made much smaller than industrial ones. The weight and size of reservoirs holding 2 to 3 minutes' flow are generally considered excessive. Although more baffling is sometimes used, air release in the smaller reservoirs is seldom complete and the practice has been to accept the increased noise.

Even smaller reservoirs are used in aircraft, so many aircraft pumps are supercharged to minimize entrained air effects. Some systems also use centrifugal separators to deaerate their fluid.

12.3.2 Suction-Line Design

Improper pump inlet lines are also a source of cavitation. Some of this is in the form of ingested air. Suction lines operate with a slight vacuum and even holes too small to leak fluid ingest air. Threaded pipe joints are notorious for such holes, so they should be avoided. There is a leakage danger with all joints, and lines should be built with a minimum of them.

The other form of suction-line cavitation is outgassing. This is the cavitation discussed in Section 8.2. It is due to reduced pressure, so suction lines should be made as short and as large in diameter as practical to minimize it. Lines should be sized to keep fluid velocity below 5 ft/sec.

Flow obstructions not only increase line pressure drop, they also cause vortices with very low pressures at their centers that spawn cavities. Lines should therefore have a minimum of joints, bends, and other discontinuities. Inlet line vacuum is also a function of pump height above the reservoir fluid surface, so this distance should be kept to a minimum. When the reservoir can be placed above the pump, it is possible to eliminate the negative pressure altogether. This is frequently done in Europe and Great Britian. It is also mandatory with high-water-base fluids to avoid vapor cavitation, which is even more violent than outgassing cavitation.

Inlet filters or even strainers are undesirable because they increase pressure drop. Filters also increase the time the fluid is exposed to vacuum, and this tends to intensify cavitation. They should therefore be located close to the reservoir where the pressure is highest. Filters that indicate when they need changing should be used to keep thc vacuum from escalating with time. Easily changed filters are preferred because they are more likely to be serviced when needed.

Occasionally, pressurized reservoirs are used to eliminate suction-line cavitation. Where pressurization is produced by gas acting directly on the fluid surface, the fluid saturation pressure rises to that of the reservoir, and cavitation is not reduced. This can be seen by comparing Figures 12.8 and 12.9. For reservoir pressurization to be effective, the fluid must not be exposed to high-pressure gas. A scheme for doing this in aircraft is shown in Figure 12.10. Systems using this method must be carefully purged of all gas and sealed before they are pressurized. Even with these precautions there may be an increase in gas absorption. However, the saturation pressure will not rise as much as the reservoir pressure, so the net effect is a reduction in cavitation potential.

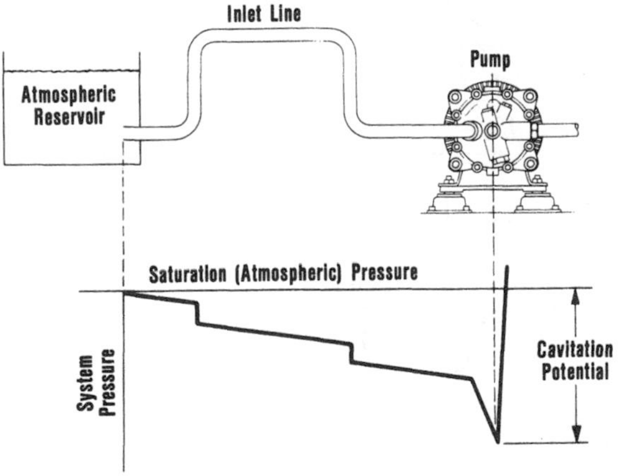

FIGURE 12.8 With an atmospheric reservoir the cavitation potential is equal to the inlet-line pressure drop. (From Ref. 5.)

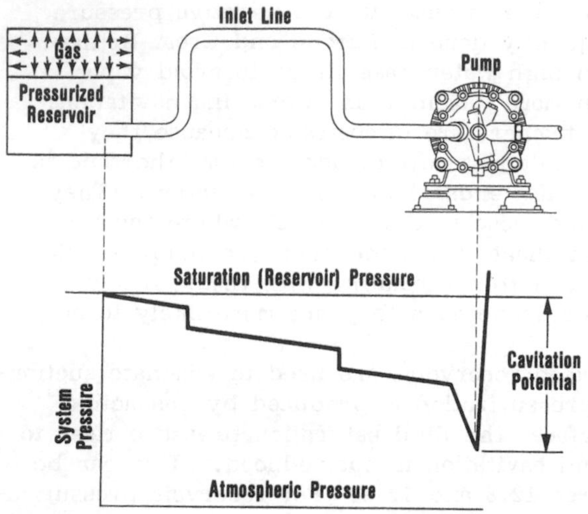

FIGURE 12.9 With a gas-pressurized reservoir the cavitation potential is still equal to the inlet-line pressure drop because the fluid becomes air saturated at the higher reservoir pressure. (From Ref. 5.)

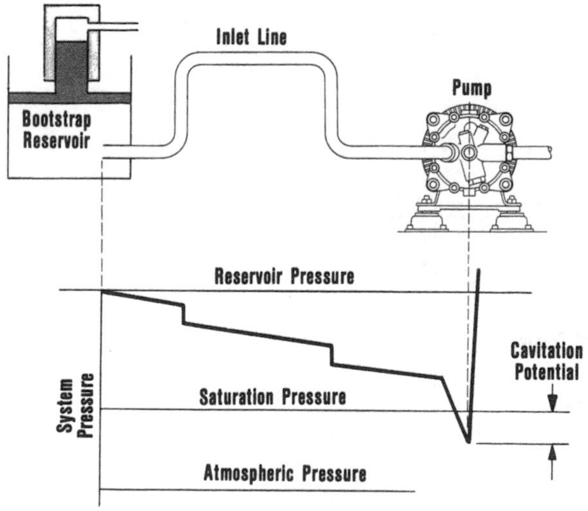

FIGURE 12.10 With a mechanically pressurized reservoir the saturation pressure does not rise as much as the reservoir pressure, so the cavitation potential is reduced. (From Ref. 5.)

12.4 FLUIDS

Fluid selection and conditioning also have a role in keeping machines quiet. The part that fluids play in hydraulic noise is often questioned.

12.4.1 Fluid Characteristics

Bulk modulus is one of the most important noise-related fluid parameters. As seen in Section 5.2.3, it is the primary fluid property entering into quiet pump port timing calculations.

The bulk modulus of common hydraulic fluids has a range of 50%. The question, then, is how much this affects pump noise. One study ran five different types of fluids in a piston pump designed to be quiet when used with petroleum oils (6). It found that fluidborne noise was much higher when both 5:95 and 60:40 oil/water emulsions were run. Structureborne noise increased with 60:40 but not 5:95 emulsions. The bulk modulus of 60:40 emulsion is the same as that of oil, while 5:95 has an 18% modulus greater modulus.

These two noises were about the same when mineral oil, phosphate ester, and water glycol were run in this pump. It is interesting to note that the bulk modulii of these fluids have a range of 50%. This suggests that bulk modulus is not critical and that the noise increases with the water emulsions is due to some factor, such as inlet cavitation. Both these fluids cavitate easily if their temperature is much above 120°F.

My experience, based on audible noise only, is that pump noise does not change with fluid as long as the pump operates within its specifications for the fluid. Limits on discharge pressure, inlet vacuum, and fluid temperature vary with the pump as well as the fluid and are usually prescribed by the pump manufacturer.

A similar observation has been made regarding viscosity. The fact that some pumps become noiser as their fluid temperature increases suggests that this was due to decreasing viscosity. Tests with different viscosity oils and temperatures show that the noise increase is related to temperature and not viscosity. It is assumed that since cavitation also increases with temperature, it is the real cause of the noise increases.

Viscosity has an indirect effect on noise, however. Increasing viscosity slows bubble release in the reservoir. It is possible therefore to increase noise by using a higher-viscosity fluid when there is not enough time in the reservoir to deaerate the oil completely before it returns to the pump.

There is one other fluid factor that affects noise. It was discovered when a hydraulic system became noisy after an oil change. It was found that the new oil did not have an antifoaming additive.

This is a chemical that is added to almost all hydraulic fluids to make bubbles burst as soon as they surface. Apparently, it also assists in bubble release, because in the subject system the new oil carried excessive entrained air to the pump.

12.4.2 Fluid Conditioning

There are two aspects of fluid conditioning that affect noise. One is fluid temperature, which like a number of other factors that have already been discussed, affects cavitation.

Quiet machines require good fluid temperature control. Optimum temperatures vary with pump specifications as well as fluid type and grade, so the pump manufacturer should be consulted for specific ranges. In general, to avoid cavitation problems, phosphate esters, water-glycols, and 60:40 oil/water emulsions should not exceed about 130°F. High-water-base fluid, often referred to as 5:95 oil/water emulsion, must be kept below 120°F for the same reason. Petroleum oils should be run at from 110 to 130°F to avoid noise problems.

Experience with lower limits for some of these fluids is scarce. However, in the case of petroleum oils, low temperatures increase viscosity. This increases suction-line pressure drop and slows bubble release in the reservoir. Machines sensitive to these high-viscosity effects indicate it with high noise levels at cold startups. The other side of the coin is that high temperature speeds up outgassing so that more will occur in an inlet line for a given vacuum.

Water in oil also causes noise. Vapor cavitation, which is more severe than gas cavitation, occurs at inlet temperatures and pressures that are satisfactory when no water is present. Condensation is a common source of water. Since it accumulates slowly, its gradual rise in noise often goes undetected. Another source is a small crack in a heat exchanger. This causes a rapid rise in noise and may be easier to catch. Water suspended in oil makes the mixture cloudy, so it is detected by routine inspections. If entrained water is neglected, it degrades pumps and valves as well as increasing their noise.

12.5 MACHINE STRUCTURE

Noise avoidance steps such as mechanical isolation and minimizing radiation efficiencies have already been dealt with. There are, however, several other concepts that should be kept in mind in designing quiet machines.

The first is that structures close to noise sources, such as motor-pump assemblies, must be kept simple. All components are possible noise radiators with tacked-on accessories being the worst offenders. These should be mechanically isolated from noise generators, or at least, located far from them. The same applies to controls such as flushing and shutoff valves that are not required for regular machine operation. Sometimes these are made less of a noise hazard by removing their handles, which are good high-frequency radiators.

The other concept is to incorporate as much mechanical isolation as cost will allow. Hose should always be used to connect pumps, motors, and valves. All hydraulic lines should be resiliently mounted. Pump and other noise source isolation mountings may cost more than the budget will permit. However, it will be considerably more expensive later, if found necessary, unless provision is made for them in the original design. The same is especially true for enclosures. Therefore, it is best to design them into the machine but leave them out of the prototype until tests determine that they are needed.

REFERENCES

1. NFPA T3.9.12, *Method of Measuring Sound Generated by Hydraulic Fluid Power Pumps*, NFPA Recommended Standard, National Fluid Power Association, Inc., Milwaukee, Wis.
2. NFPA T3.9.14, *Method of Measuring Sound Generated by Hydraulic Fluid Power Motors*, NFPA Recommended Standard, National Fluid Power Association, Inc., Milwaukee, Wis.
3. *NMTBA Noise Measurement Techniques*, 2nd ed., National Machine Tool Builders Association, McLean, Va.
4. *More Sound Advice*, Vickers Inc., Troy, Mich.
5. *Quiet Please*, Vickers Inc., Troy, Mich.
6. J. Kelsey, R. Taylor, and K. Foster, "Fluid Properties: The Effect of the Fluid Being Pumped on the Noise Emitted by an Axial Piston Pump," paper C384/80, *Proceedings of the Quieter Oil Hydraulics Seminar*, Institution of Mechanical Engineers, Oct. 1980, pp. 71–75.

13
Diagnosing Noise Problems

Noise measurements are the key to efficient noise reductions. They define the problem, identify secondary radiators, detect resonances, and suggest noise energy paths. At the completion of a project they confirm the results.

It is helpful to think of a machine as having many different noises, even when there is only one source present. Since noise transmission is sensitive to frequency, energy at the various pump noise component frequencies, for example, has differing responses to noise control measures. Similarly, energy has several different paths from its source to a listener, and each path affords different control opportunities. Each of these subdivisions, then, has unique behavior and control options, so they should be treated as separate noises.

We know that a machine's overall noise level is not significantly reduced until the strongest of these noises are attenuated. The first objective of diagnostic testing is to identify the few critical noises. Tests are then used to indicate their source, where they are radiated, and their path between the source and radiator. When some of these facts are known, remedies are usually obvious. Because of this, noise reduction is largely a job of diagnostic testing.

We seldom learn these things with certainty, so it is necessary to try a number of remedies suggested by the tests. This, in itself, is a form of diagnostic testing because each trial either confirms or disproves a tentative hypothesis drawn from earlier test results.

This chapter is directed to quieting machines because it is a broader subject than reducing component noise. Techniques appropriate to components are covered. However, because components are relatively compact, some of the analyses discussed are not useful when working with them.

13.1 IMPLEMENTATION

Precision measurements made in acoustically good sound rooms with good-quality instruments, although preferred, are not essential for diagnostic testing. Tests made in regular work spaces with inexpensive sound-level meters usually provide adequate information for quieting a machine.

13.1.1 Instrumentation

Diagnostic testing requires measuring in many locations. Portability and fast performance are more important than precision. Most measurements are not recorded, so instruments with graphic or numerical printouts only impede this type of work.

The sound-level meter used in diagnostic testing must provide octave band as well as A-weighted levels. The filters needed to measure these levels were discussed in Section 4.4.1. Third-octave filters are desirable but not needed very often, so it is best to have an instrument that permits choosing the band width. It is not a big handicap if the budget cannot be stretched enough to buy an instrument with third-octave filters.

Simple hand-held sound-level meters are capable of providing adequate accuracy and are preferred in examining large machines. Their utility is increased by the ability to connect the microphone through a cable. Similarly, the diagnostic capability of the meter is greatly enhanced if it can be used with accelerometers.

Discrete frequency analyzers are sometimes useful in screening for resonances and for source identification. Their operation is too slow for use in surveys, so their lack of portability does not cause problems. They should have a microphone cable long enough to permit measurements anywhere around a machine, however. Even this is not necessary if noises can be tape recorded and then analyzed. This alternative is especially useful for mobile machine studies.

13.1.2 Test Space

Special noise test rooms are seldom available for machine studies. Most machines require very large, semianechoic facilities for precision measurements. Fortunately, diagnosis does not depend on such accuracy and most tests are made successfully in ordinary work spaces. Acoustically, all that is required is to locate the machine away from large reflecting surfaces. Frequently, such locations are not free of noise interference during regular working hours, but good data are acquired during lunch hours and at night.

13.1.3 Controls

Some diagnostic testing is in the form of measuring the effect of modifications. This includes changes made to test a hypothesis about a noise as well as those made as "fixes," so errors can cause the investigation to go off in the wrong direction. Since evaluations are the difference in two readings, consistency is more important than measurement accuracy.

Consistency depends on the repeatability of machine rather than instrument operations. The need for good pressure, vacuum, and temperature gauges and controls is often overlooked. Motor speeds are affected by other electrical loads on the same line, so they should be monitored to make sure that the speeds are the same for both measurements.

Unusual controls may also be needed. For example, in some systems the dissolved air changes with operating time, so they must be run for a long period to stabilize them. Similarly, closed-loop system lines must be carefully bled during reassembly because the noise of such systems is usually sensitive to residual air pockets.

13.2 NOISE-LEVEL MEASUREMENTS

All noise reduction programs must start by measuring the noise level with which you are dealing. The first objective is to find how much noise reduction is needed. This is particularly true with projects that are initiated because of subjective judgments or customer complaints.

Good numbers are required. Since most noise reductions are OSHA related, noise levels are measured at the operator's position and other places around the machine where workers may be exposed to the machine noise. This is done as described in Section 4.4.4 but measurements are not limited to standard locations; special efforts must be made to find the highest level to which people will be exposed. For purposes of thus discussion, this highest level is called the key level to distinguish it from higher levels within the machine, which are not a concern at this stage of an analysis.

As discussed in Section 12.1.2, the amount of noise reduction needed determines how much effort is required to achieve it. It provides direction as well. When a modest reduction is needed, minor changes can provide it, so major modifications are not even considered. The reverse is true when a large reduction is needed.

The diagnostic effort starts with these noise-level measurements. They find where noise is highest and what event in the machine's work cycle causes the most noise. Testing effort is minimized by making all following tests at only the worst operating conditions and measuring, mostly, in locations where the strongest levels occur.

13.3 FREQUENCY ANALYSES

Frequency analysis is our principal diagnostic tool. We use it for examining the key noise to find clues to its source and some things about how it is propagated.

13.3.1 Source Identification

When a hydraulic pump is known to be the noise source, the noise frequencies are found from its speed and number of pumping chambers. All that needs to be measured is the strengths of its harmonics. Generally, octave-band analyses provide this information adequately even though some bands include more than one harmonic.

Pumps are relatively small in comparison to the wavelengths of their first few pumping harmonics, so their radiation at these frequencies is inefficient. Although excitations at these frequencies are strong, a higher harmonic that is efficiently radiated predominates the pump airborne noise spectrum. If fluid or structureborne noise energy excites some radiator larger than the pump, radiation efficiency of the lower harmonics is increased. Airborne levels of these harmonics are then raised relative to the pump's peak harmonic and some may exceed it. Comparing the spectra shapes of the key and pump noises, then, sometimes tells whether this noise is from a large radiator.

When there is more than one noise source, frequency analyses are used to identify which causes the loudest sound. No frequency analyses, of course, can separate noise sources having similar spectra, such as pumps with the same pumping frequency. They can be used to separate the effects of pumps having different pumping frequencies, pumps and motors running at different speeds, or nonhydraulic noise sources coexisting with a pump. Octave band analyses may be too broad for this task, and one-third-octave measurements may be needed to separate harmonic frequencies of different sources. With complex machines such as agricultural combines, discrete frequency analyses may be required to pinpoint the critical noise source.

13.3.2 Resonances

Frequency analyses also help identify resonances. When these cause excess noise, retuning the resonator dramatically reduces this noise, When a strong resonance exists, its level overshadows all others and is easily found from discrete frequency spectra and sometimes from one-third-octave analyses. It is harder to spot in octave-band data. Another way to detect an important resonance is to examine the microphone signal with an oscilliscope. When a resonance is present, thc signal is nearly a pure sine wave.

The surest way to detect a resonance is to make a small change in the speed of the noise source. If the resonance is a strong one, the change will have a large effect on the noise level of the predominant harmónic. Unfortunately, such speed changes are seldom possible.

13.4 IDENTIFYING RADIATORS

The search for strong radiators should focus on the highest-octave or third-octave band in the key noise spectrum. This is the band that must be reduced to make the machine quieter. Ignoring the others not only speeds up the process, it also reduces the chances of overlooking important information. If two bands have nearly the same high level, it is best to make separate surveys for each rather than try to survey for both at the same time.

13.4.1 Microphone Surveys

The simplest way to search for powerful noise radiators is to move the microphone around the machine to find "hot spots." A hand-held meter is best for this search. The person making it can see how the level rises and falls as they move the microphone and so can quickly home in on high-level locations. The microphone should be cable connected so that the meter is readable regardless of the microphone position.

The objective is to find the surfaces radiating the highest levels. However, levels drop off rapidly as the microphone moves away from a radiator, so measurements must be made at a constant distance from the various surfaces. A toothpick attached to the microphone with a rubber band, projecting about an inch in front of the microphone face, is handy for gauging this distance.

13.4.2 Accelerometer Surveys

The search can also be made with an accelerometer. Results above 1000 Hz are uncertain, however, because of difficulties in maintaining good contact with surfaces. The transducer must be held tightly against surfaces and this affects sound from sheet metal panels. It does not affect noise from thicker metal such as structural shapes or mounting plates.

The accelerometer can be attached to flat iron or steel surfaces with a magnet. These are sold as accelerometer accessories and they free the operator's hands while permitting the accelerometer to be moved from place to place. This setup often cannot be used on thin sheet metal panels, however, because the mass of the magnet and accelerometer changes their resonant frequencies.

13.4.3 Cladding

As discussed in Section 4.1.7, because of cancellation, a quadrupole noise source can produce high near-field levels without radiating much to the far field. Survey techniques that measure near-field levels or surface deflections, therefore, find only potentially critical radiators. Often cut-and-try machine modifications are used to determine which are really critical.

A technique for positively identifying powerful radiators is to cover suspected surfaces and measure the effect. This method is too cumbersome to be used for surveying, but it is more definitive than survey techniques.

Cladding, the flexible barrier material discussed in Section 9.3.5, is used. Great care is required in applying it and in sealing around its edges. The effect of the radiation from the subject surface is measured directly by monitoring the highest-octave or third-octave level at the key noise location.

Appendix A
SI Units

SI units are from the meter-kilogram-second system, in which the units are often named after famous scientists. Abbreviations for units named after people are capitalized and may consist of either one or two letters. The units themselves are not capitalized. For example, the unit for pressure is the *pascal*, Pa.

SI units commonly encountered in noise reduction work are listed below with their English unit equivalents.

SI UNITS

Quantity	Unit	Equivalent	English equivalent
Length	meter	m	39.4 in. 3.28 ft
Frequency	hertz	Hz	cps
Mass[a]	kilogram	kg	0.0686 lb-sec^2/ft[b] 2.20 lb (weight)[c]
Density		kg/m^3	0.00194 lb-sec^2/ft^4
Force	newton	N = kg-m/sec^2	0.225 lb
Pressure	pascal	Pa = N/m^2	1.45 × 10^{-4} psi
Energy (work)	joule	J = N-m	0.738 lb-ft
Power	watt	W = J/sec, 745 W	0.738 lb-ft/sec, 1 hp

Quantity	Unit	Equivalent	English equivalent
Impedance (acoustic)		Z = N-sec/m	2.37×10^{-9} lb-sec/in.^{5}

[a]The English system does not have a widely accepted unit for mass.
[b]This is the "slug" used in older mechanics books; one weighs 32.2 lb.
[c]The unit lb (weight) is not consistent with other units and is confused with lb, the force unit.

Prefixes are added to the abbreviations to indicate decimal multiples and submultiples. The ones commonly used in noise work are listed below. With the exception of mega, these are never capitalized, even though they are the first letter.

UNIT PREFIXES		
M	mega	10^6
k	kilo	10^3
d	deci	10^{-1}
c	centi	10^{-2}
m	milli	10^{-3}
μ	micro	10^{-6}
n	nano	10^{-9}
p	pico	10^{-12}

All logarithms used in sound and noise control are to the base 10. Some of the common constants used in this work are listed below.

COMMON CONSTANTS

	SI	English
g acceleration due to gravity	9.81 m/sec^2	32.2 ft/sec^2 386 in./sec^2
bar atmospheric pressure	0.751 m Hg 10^5 Pa	29.6 in. Hg 14.5 psi
ρ[a] air density	1.18 kg/m^3	0.00229 lb-sec^2/ft^4
c[a] speed of sound	344 m/sec	1130 ft/sec
ρc[a] characteristic impedance	406 N-sec/m^3	2.59 lb-sec/ft^2

[a]Air at standard conditions of 22°C (71.6°F) and 0.751 m (29.6 in.) Hg.

Appendix B
Noise-Level Ratios

Sound strengths are measured in terms of decibels, which are proportional to logarithms to the base of 10. Scales are provided for both pressure and power levels. These are defined as follows:

sound power level

$$L_w = 10 \log \frac{W}{W_0} \quad \text{dB}$$

where

W = sound power, W
W_0 = reference sound power, 10^{-12} W

sound pressure level

$$L_p = 10 \log \frac{P^2}{P_0^2}$$

$$= 20 \log \frac{P}{P_0} \quad \text{dB}$$

where

P = rms sound pressure, Pa
P_0 = rms reference pressure, 20 μPa

Comparisons of sound powers and pressures are made by taking the difference in their decibel levels. Since the decibel scale is logarithmic, this has the effect of dividing the higher one's strength by the lower one's strength. The difference represents their strength ratio, which is given in the following table.

Values exceeding those covered by the table are reduced to the table range by subtracting multiples of 10 for power ratios and multiples of 20 for pressure ratios. The ratio corresponding to the reduced difference is then multiplied by the appropriate factor given in the second table below.

NOISE-LEVEL RATIOS

Difference (dB)	Pressure ratio	Power ratio	Difference (dB)	Pressure ratio
0.5	1.06	1.1	10.5	3.4
1.	1.1	1.3	11.	3.6
1.5	1.2	1.4	11.5	3.8
2.	1.3	1.6	12.	4.0
2.5	1.3	1.8	12.5	4.2
3.	1.4	2.0	13.	4.5
3.5	1.5	2.2	13.5	4.7
4.	1.6	2.5	14.	5.0
4.5	1.7	2.8	14.5	5.3
5.	1.8	3.2	15.	5.6
5.5	1.9	3.6	15.5	6.0
6.	2.0	4.0	16.	6.3
6.5	2.1	4.5	16.5	6.7
7.	2.2	5.0	17	7.1
7.5	2.4	5.6	17.5	7.5
8.	2.5	6.3	18.	7.9
8.5	2.7	7.1	18.5	8.4
9.	2.8	7.9	19.	8.9
9.5	3.0	8.9	19.5	9.4
10.	3.2	10.0	20.	10.0

RATIO MULTIPLIERS

Decibels subtracted from difference	Multipliers: Pressure ratio	Multipliers: Power ratio
10		10
20	10	10^2
30		10^3
40	10^2	10^4
50		10^5
60	10^3	10^6

Example 1

$$\begin{aligned} L_1 &= 87 \text{ dB} \\ L_2 &= 79 \\ \hline L_1 - L_2 &= 8 \text{ dB} \end{aligned}$$

pressure ratio = 2.5
power ratio = 6.3

Example 2

$$\begin{aligned} L_3 &= 94 \text{ dB} \\ L_4 &= 61 \\ \hline L_3 - L_4 &= 33 \text{ dB} \end{aligned}$$

Pressure:

$33 - 20 = 13$

pressure ratio = $4.5 \times 10 = 45$

Power:

$33 - 30 = 3$

power ratio = $2.0 \times 1000 = 2000$

Index

Milton Keynes UK
Ingram Content Group UK Ltd.
UKHW021626071024
449327UK00020BA/1204

9 780367 451295